T0269295

DSP 320F28335 Programming

Published 2024 by River Publishers

River Publishers

Alsbjergvej 10, 9260 Gistrup, Denmark

www.riverpublishers.com

Distributed exclusively by Routledge

605 Third Avenue, New York, NY 10017, USA

4 Park Square, Milton Park, Abingdon, Oxon OX14 4RN

DSP 320F28335 Programming / by Majid Pakdel.

Routledge is an imprint of the Taylor & Francis Group, an informa business

ISBN 978-87-7004-197-3 (paperback)

ISBN 978-87-7004-218-5 (online)

ISBN 978-8-770-04211-6 (ebook master)

A Publication in the River Publishers Series in Rapids

While every effort is made to provide dependable information, the publisher, authors, and editors cannot be held responsible for any errors or omissions.

DSP 320F28335 Programming

Majid Pakdel

MISCO Company, Mianeh, Iran

River Publishers

Routledge
Taylor & Francis Group
NEW YORK AND LONDON

Contents

Preface

This book covers the programming of the popular Texas Instruments C2000 microcontroller family, specifically the TMS320F28335, which is widely used in implementing digital signal processing (DSP). In the first chapter, an introduction to microprocessors is provided. The second chapter delves into memory organization and I/O ports. The third chapter discusses system interrupts and timers. The fourth chapter explains digital-to-analog converters, and the fifth chapter elucidates analog-to-digital converters, and finally, in the sixth chapter, code generation using simulation software such as Simulink, PSIM, and PLECS is explored. This book serves as an excellent guide for hardware implementation of control algorithms and DSP for electrical and computer engineering students.

I dedicate this book to the loving memory of my mother, Rahimeh Ghorbani, with a great and kind soul who dedicated her entire blessed life to education and love for her children. It is regret that I did not fully appreciate her while she was alive, and only after her passing did, I realize what a precious gem she was. Unfortunately, I lost her, and nothing in the world can replace her. I hope she will forgive me, and I pray that her soul rests in eternal peace.

About the Author

Majid Pakdel received his Bachelor's degree in Electrical-Telecommunications Engineering from Amirkabir University of Technology, Tehran in 2004, his Master's degree in Electrical power Engineering from Isfahan University of Technology in 2007, his Ph.D. in Electrical power Engineering from University of Zanjan in 2018, and his Master's degree in Computer Engineering-Artificial Intelligence and Robotics from Malek Ashtar University of Technology, Tehran in 2023. He has published over 20 papers and 3 books in the fields of electrical engineering and computer science. He was a visiting Ph.D. student in the Department of Energy Technology at Aalborg University between 2015 and 2016.

1

Introduction to Microprocessors

1.1 Introduction

The structure of this book is as follows.

Chapter 1

- General presentation of microprocessors
- Human interface: code composer studio.

Chapter 2

- Memory organization
- I/O ports.

Chapter 3

- System interrupts
- Timers.

Chapter 4

- Digital to analog conversion.

Chapter 5

- Analogue to digital conversion.

Chapter 6

| •   Code generation using Simulink, PSIM and PLECS.

1.2 General Presentation of Microprocessors

In this book we will use the C programming language which includes the following features:

| •   Variables
| •   Functions
| •   Pointers
| •   Conditional statements (IF-ELSE)
| •   Loops: FOR, While, DO-WHILE.

We will do coding while learning it throughout the book and do coding for a project in a real job. The application examples of microcontrollers (µC) are as follows.

| •   Industrial:

| o   Process control, robotics, atomic reactors, wind turbines, etc.

| •   Home usage:

| o   Washing machines, microwaves, etc.

| •   Automotive:

| o   ABS, ESP, engine control etc.

| •   And many more applications.

What is a microprocessor (µP)? Microprocessors contain only a central processing unit (CPU) and need to connect to external memory and peripherals to be able to operate, as shown in Figure 1.1.

The computer peripherals include the following features:

| •   Digital input/output lines (ex. parallel port)
| •   Analogue to digital converter (ADC) (ex. microphone in)

Figure 1.1: The microprocessor unit with memory and peripherals.

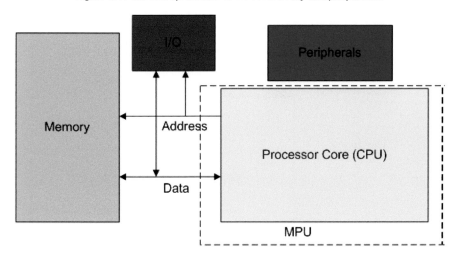

- Digital to analogue converter (DAC) (ex. audio output)
- Timer/counter units
- Network interface units:

  ○ Serial interface (RS232, USB)
  ○ Computer networking (Ethernet)

- Graphical Output Devices
- And more....

The microcontroller (μC) is nothing more than a "PC + peripherals" on a single silicon chip or a system on chip and it has the following characteristics:

- All computing power AND input/output channels that are required to design a real time control system are "on chip".
- Guaranteed cost efficient and powerful solutions for the embedded control applications.
- Backbone for almost every type of modern product.
- Over 200 independent families of μC.
- Both μP-architectures ("Von Neumann" and "Harvard") are used inside microcontrollers.

What is a microcontroller (MCU)? In the microcontrollers the CPU, memory and peripherals are integrated into the chip as illustrated in Figure 1.2.

Figure 1.2: The microcontroller unit (MCU).

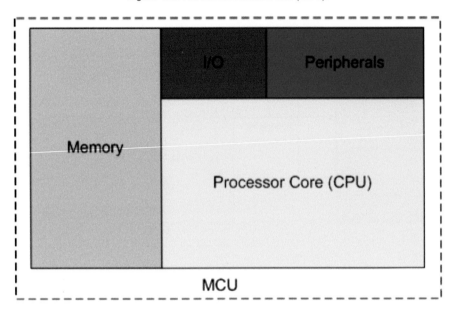

What is real-time control? Steps of a typical control algorithm are as follows:

- Data acquisition (measurement of the controlled parameter like temperature, voltage speed etc.).
- Calculation of the new control values (voltage or current vector, position etc.).
- Load the control values in the peripherals (PWM, D/A units).

In summary, a fast response based on the measurements of the environmental parameters is called "real-time control". An example of a real-time application is to maintain constant temperature of an oven using DSP as shown in Figure 1.3.

The "Von Neumann" microprocessor architecture is shown in Figure 1.4. This structure has the pros and cons; the main advantage of it is that it has easier memory management. However, its drawback is that it is slower, since the instruction and data have to be fetched separately. Therefore, in Neumann architecture, reading of an instruction or reading/writing of a data from the memory cannot be in the same time.

The "Harvard" microprocessor architecture is depicted in Figure 1.5.

Figure 1.3: The real-time control of an oven using DSP.

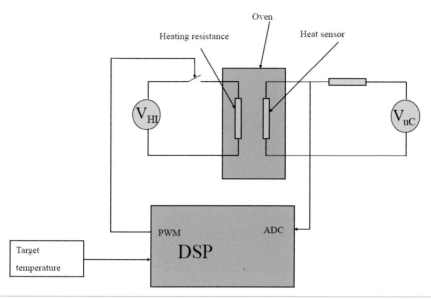

Figure 1.4: The Neumann architecture.

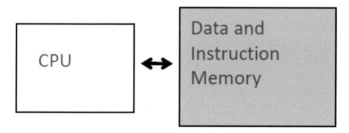

The main advantage of the Harvard architecture is that it is faster and it is used as the usual architecture for a DSP. The main drawback of this structure is that its memory management is more complex. Therefore, in Harvard architecture, reading of an instruction or reading/writing of a data from the memory can be in the same time.

What is a digital signal processor (DSP)? It has the following characteristics:

- Similar to a microprocessor (μP).
- Additional hardware units to speed up computing of sophisticated mathematical operations:

Figure 1.5: The Harvard architecture.

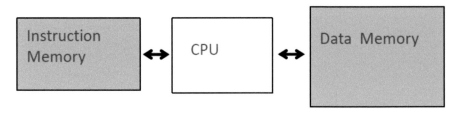

- o   Additional hardware multiply unit(s)
- o   Additional floating-point unit
- o   Additional pointer arithmetic unit(s)
- o   Additional bus systems for parallel access
- o   Additional hardware shifter for scaling and/or multiply/divide by 2^n.

- Usually used and has limited number of peripherals (input/output ports).
- It is used for sophisticated mathematical calculations.

What is a digital signal controller (DSC)? It has the following characteristics:

- Recall: A microcontroller (μC) is a single chip microcomputer with a microprocessor (μP) as core unit.
- Now: A digital signal controller (DSC) is a single chip microcomputer with a digital signal processor (DSP) as a core unit.
- By combining the computing power of a DSP with memory and peripherals in one single device we derive the most effective solution for embedded real time control solutions where sophisticated mathematical calculations are required.
- DSC. Example: Texas Instruments C2000 family.
- It is very common to call a DSC a DSP.

The sum of products (SOP) is the key element in most DSC algorithms. The typical DSP algorithms are illustrated in Figure 1.6.

How will an SOP look in a DSP? A C-code solution could for $y = \sum_{i=0}^{3} \text{data}[i] \times \text{coeff}[i]$ look like this:

#include <stdio.h>

intdata[4]={1,2,3,4}, coeff[4]={8,6,4,2};

intmain(void){

Figure 1.6: The typical DSP algorithms.

Algorithm	Equation
Finite Impulse Response Filter	$y(n)=\sum_{k=0}^{M} a_k\, x(n-k)$
Infinite Impulse Response Filter	$y(n)=\sum_{k=0}^{M} a_k x(n-k)+\sum_{k=1}^{N} b_k y(n-k)$
Convolution	$y(n)=\sum_{k=0}^{N} x(k)h(n-k)$
Discrete Fourier Transform	$X(k) = \sum_{n=0}^{N-1} x(n)\exp[-j(2\pi/N)nk]$
Discrete Cosine Transform	$F(u)= \sum_{x=0}^{N-1} c(u).f(x).\cos\left[\dfrac{\pi}{2N}u(2x+1)\right]$

```
inti;

intresult =0;

for (i=0;i<4;i++)

result += data[i]*coeff[i];

printf("%i",result);

return 0;

}
```

What will a processor be forced to do to solve an SOP, $y = \sum_{i=0}^{3} \text{data}[i] \times \text{coeff}[i]$? The basic operations of an SOP are as follows:

1. Set a Pointer1 to point to data[0]
2. Set a second Pointer2 to point to coeff[0]
3. Read data[i] into core
4. Read coeff[i] into core

5. Multiply data[i]*coeff[i]
6. Add the latest product to the previous ones
7. Modify Pointer1
8. Modify Pointer2
9. Increment I
10. If i < 3, then go back to step 3 and continue.

The steps 3 to 8 are called the "six basic operations of a DSP". A DSP is able to execute all 6 steps in one single machine cycle!

The SOP machine code of a DSP is shown in Figure 1.7.

Figure 1.7: The SOP machine code of a DSP.

Address	M-Code	Assembly - Instruction	
10:	for (i=0;i<4;i++)		
00411960	C7 45 FC 00 00 00 00	mov	dword ptr [i],0
00411967	EB 09	jmp	main+22h (411972h)
00411969	8B 45 FC	mov	eax,dword ptr [i]
0041196C	83 C0 01	add	eax,1
0041196F	89 45 FC	mov	dword ptr [i],eax
00411972	83 7D FC 04	cmp	dword ptr [i],4
00411976	7D 1F	jge	main+47h (411997h)
...			

What is a fixed-point DSC? The decimal numbers have an integer and a fractional part which can be represented by using a 16-bit integer variable, as shown in Figure 1.8.

The "IQ math" library can be used for conversion and mathematical operations between different "Q" classes.

What is a floating-point DSC? It has the following characteristics:

- The floating point number can be represented as:

  ○ where b-base, s-significand and e-exponent

$$s \times b^e.$$

- The floating point unit is an extra unit in DSP which speeds up the calculations with floating numbers

Figure 1.8: The fixed-point DSC.

10011101.01010011 or 10.01110101010011
Integer
fractional

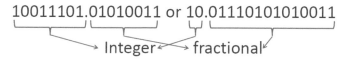

For example:

float x=0.03, y=123.88, Z;

Z = y/x;

It takes around 100 assembly instructions in a fixed point DSP and takes around 8 assembly instructions in a floating-point DSP. In a fixed-point DSP, by usage of "IQ math", the same performance can be achieved, but requires software development overhead. The DSP 320F28335 experimental kit is depicted in Figure 1.9.

Figure 1.9: The DSP 320F28335 experimental kit.

DSP 320F28335
JTAG interface
USB Interface
On/Off switch
Power connector
Analog Interface
I/O Interface

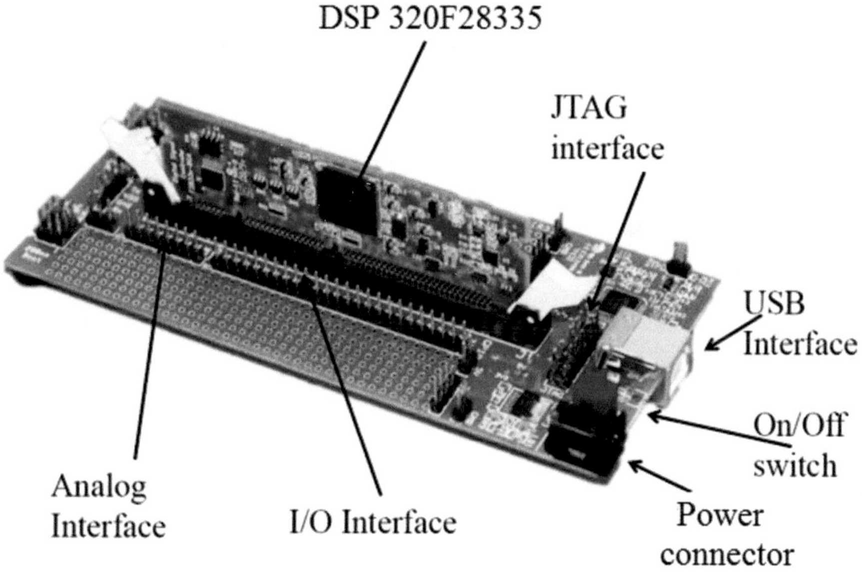

1.3 Human Interface: Code Composer Studio

In connecting the experimental kit to your PC, we have the following features:

- The code composer studio–software (SW) interface on the PC
- The board can be powered through USB, there is the possibility of external power.

The code composer studio includes the following features:

- Supported software languages: Assembly, C, C++.
- A desktop PC is used for DSP algorithm development, linked with DSP through a USB cable.
- Code simulation.
- Real-time emulation (usually through a JTAG interface).

A block diagram of the code composer studio is illustrated in Figure 1.10.

Figure 1.10: A block diagram of the Code Composer Studio.

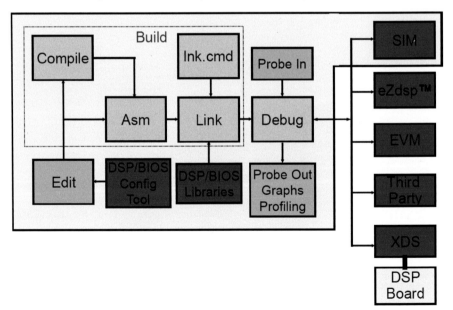

The different parts in the Code Composer Studio® IDE are shown in Figure 1.11.

Figure 1.11: The different parts of the Code Composer Studio® IDE.

Project Manager:
➢Source & object files
➢File dependencies
➢Compiler, Assembler & Linker build options

Menus or Icons

Full C/C++ & Assembly Debugging:
➢C & ASM Source
➢Mixed mode
➢Disassembly (patch)
➢Set Break Points
➢Set probe Points

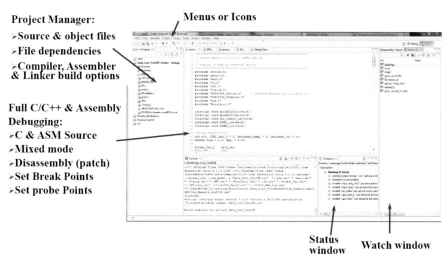

Status window   **Watch window**

The CCS project (.pjt) files contain:

- Source files (by reference)

    o Source (C, C++, and assembly)
    o Libraries
    o DSP/BIOS' configuration
    o Linker command files.

- Project settings:

    o Build options (compiler and assembler)
    o Build configurations
    o DSP/BIOS
    o Linker

The C/C++ compiler includes the following features:

- Translates the C/C++ program code to machine code (assembly code).
- The translated code can be stored .lib files.

    o Library is a collection of subroutines or classes used to develop software.

11

- The subroutines are includes in C code through header files.
- The subroutines can be stored in C files, in this case it is translated to machine code on every time the project is built.

The general extension language (GEL), gives the possibility of creation of a GUI for real-time control in the following formats:

- Slider
- Dialog.

For an example consider the following code:

menuitem "Mot_Speed"

slider Mot_speed(0,1500,1 ,1,Mot_speed){

Speed = Mot_speed;

}

The slider range is from 0 to 1500 and it is incremented and decremented by 1 and "Speed" is the software (SW) variable. The "Mot_speed" slider can be accessed from the "Scripts" tab in menus as you can see in Figure 1.12 and Figure 1.13 respectively.

Figure 1.12: The "Mot_speed" slider in the "Scripts" tab.

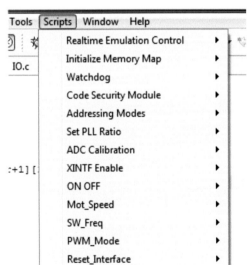

Figure 1.13: The "Mot_speed" slider.

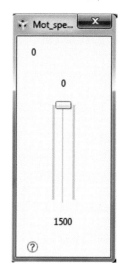

In summary, we have examined the following topics in this chapter:

- General presentation of microprocessors:

    ○ Structure of microprocessors
    ○ Types: microprocessor, microcontroller, DSP, DSC
    ○ Peripherals.

- Code Composer Studio.

1.4 Lab1

During the laboratory exercises we will build up step by step a project, which means that we start by creating a new project from zero and we will add to this project new files which contains the control code. Lab1 is a starts with F28335 DSP which includes the following experiments:

- Create a new project.
- Write a program which you can do:

    ○ Basic arithmetic operation
    ○ Set the value of a variable using GEL slider.

13

In this lab, you will perform the initialization of the DSC and you will make basic arithmetic operations using the DSP board. Although obviously simplistic, it is good to see how to create a project step by step, and program the DSP board for the first time. The creation of a project in Code Compose Studio is as follows:

(1) Ensure that the DSP board is connected to the PC, and powered up.

(2) Run Code Composer Studio 6.x.

(3) Press the fan "Project" and chose the "New CCS Project" command and fill:

  • Project name: MicroprocessorsLab
  • For location: you can use your own folders for this
  • Target: 2833x Delfino → Experimenter's Kit - Delfino F28335

(4) From "Project templates and examples" choose Empty Project (with main.c), as shown in Figure 1.14.

(5) Choose the "Empty Project" and click "Finish".

Figure 1.14: Choosing Empty Project (with main.c).

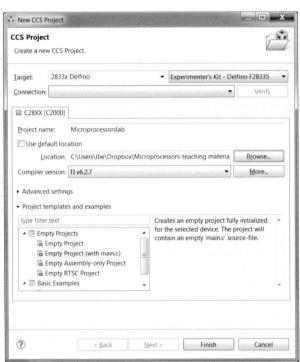

Link the source files using right click on "MicroprocessorsLab" and selecting the "Add Files…" command. When you have chosen one file or more, press open. A dialog will then popup where you can choose "Link to files". This has to be done for all the files. Link the following files, which can be found in the ti/controSUITE folder (the controlSUITE has to be installed previously):

- DSP2833x_Headers_nonBIOS.cmd from
  c:\ti\controlSUITE\device support\f2833x\v140\DSP2833x_headers\cmd\
- DSP2833x_GlobalVariableDefs.c from
  c:\ti\controlSUITE\device support\f2833x\v140\DSP2833x_headers\source\
- DSP2833x_CodeStartBranch.asm, DSP2833x_ADC_cal.asm,
  DSP2833x_PieCtrl.c, DSP2833x_PieVect.c, DSP2833x_SysCtrl.c, DSP2833x_DefaultIsr.c,
  DSP2833x_usDelay.asm from
  c:\ c:\ti\controlSUITE\device support\f2833x\v140\ DSP2833x_common\source\

The project directory should look like that depicted in Figure 1.15.

Figure 1.15: The project directory.

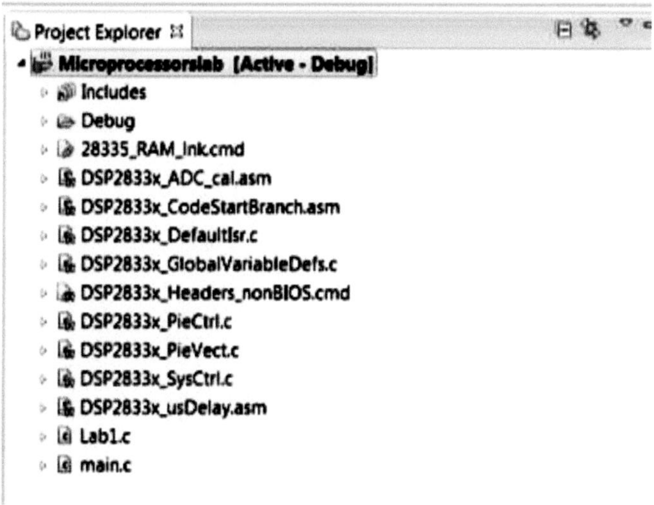

(6) Set the Path for the header files using again right click on "MicroprocessorsLab" selecting "Properties" command. Under the fan "C2000 Compiler", "Include Options" and in the second area add the following two as illustrated in Figure 1.16:
- c:\ti\controlSUITE\device support\f2833x\v140\DSP2833x_headers \include\
- c:\ti\controlSUITE\device support\f2833x\v140\DSP2833x_common\ include\

Figure 1.16: Adding two lines in "Include Options".

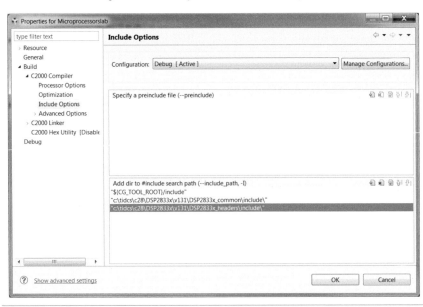

(7)   Create a new file using File\New\Source File with the name Lab1.c.

(8)   When the files open, write the following source code into the file:

```
#include "DSP28x_Project.h" // Device Headerfile and Examples Include File
int i=2,j=3,k;
main(){
// Initialize System Control: PLL, WatchDog, enable Peripheral Clocks
InitSysCtrl();
// Disable CPU interrupts
DINT;
// Initialize PIE control registers to their default state.
InitPieCtrl();
// Disable CPU interrupts and clear all CPU interrupt flags:
IER = 0x0000;
IFR = 0x0000;
// Initialize the PIE vector table with pointers to the shell Interrupt
InitPieVectTable();
while(1){
k=i+j;
}
}
```

Then, do the following steps:

(1) Build the code by right click on "MicroprocessorsLab" and select "Build Project".

(2) Right click on the "MicroprocessorsLab" directory and select New/Target Configuration File.

(3) Create the new target configuration file with the name as below: NewTargetConfiguration.ccxml

(4) Select the experimental kit as shown in Figure 1.17.

Figure 1.17: Selecting the experimental kit.

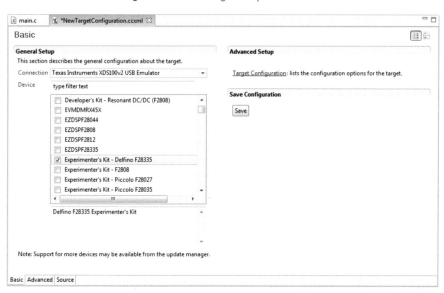

(5) Load the code in the DSC by clicking on the green bug pictogram, as depicted in Figure 1.18.

Figure 1.18: Clicking on the green bug pictogram.

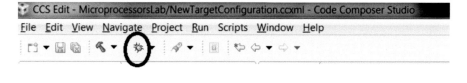

Congratulation, you made your first project in CCS!

For testing your first real code do the following steps:

(1) Select variable i, j, and k one by one by dragging them to the "Expressions" area, or by copying the name into the area.
(2) By using this you can perform the arithmetical operation and see the result on the Watch Window.
(3) On the watch window the variables should have the given values, in the "k" variable the sum should appear.
(4) Now press "Reset CPU" and after "Restart" to restart the execution of your code as illustrated in Figure 1.19.

Figure 1.19: Pressing "Reset CPU" and after "Restart".

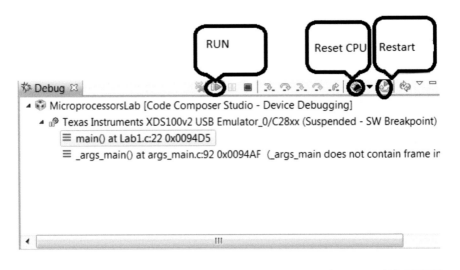

(5) Place a Breakpoint in the Lab1.c – window at line "k = i + j;" Do this by placing the cursor on this line, click right mouse and select: "Breakpoint" and then "Breakpoint" again. The line is marked with a blue dot to show an active breakpoint (if it is gray means, it is not active). Perform a real-time run by pressing F8. The program will stop execution when it reaches the active breakpoint.
(6) By changing the value of the variables i and j you can change the value of k.

For using a slider do the following steps:

(1) Create a new file using File\New\File with the name Input_Val.gel
(2) Place the next cod in the file:

menuitem "Input_Val"

slider Input_Val (0,50,1 ,1, Input_Val){

Input= Input_Val;

}

(3) Save the file as slider.gel in the Lab project directory.
(4) Go back to the Lab1.c project file and add the next code:

```
#include "DSP28x_Project.h" // Device Headerfile and Examples Include //File
int i=2,j=3,k;
int Input; //add this
float d = 2.2, e, f; //add this
void main(){
// Initialize System Control: PLL, WatchDog, enable Peripheral Clocks
InitSysCtrl();
// Disable CPU interrupts
DINT;
// Initialize PIE control registers to their default state.
InitPieCtrl();
// Disable CPU interrupts and clear all CPU interrupt flags:
IER = 0x0000;
IFR = 0x0000;
// Initialize the PIE vector table with pointers to the shell Interrupt
InitPieVectTable();
while(1) {
e=((float) Input) / d; //add this
}
}
```

(5) Load the gel file by clicking on Tools/Gel Files (new window will appear nearby the Watch window), right click on the line below f28335.gel and load the created GEL file (make sure that you are in "CCS Debug").

In the Script menu a new tag will appear with the name: "Input_Val". By clicking on it, a slider window is going to appear.

(6) Build and run the code.

(7) Add the variable "e" to the watch window.

(8) By moving the position of the slider, the value of variable "e" will have the value of the floating-point arithmetical calculation.

(9) Your task is to write a function which calculates the roots of a second order function a*x∧2+b*x+c (in case there are imaginary roots showing the root as zero). The function should have three floating point input parameters: a, b, c. the roots should be global variables for ex. x1 and x2.

(10) Use a slider for each coefficient (a, b, c) in order to modify the parameters of the second order function.

(11) Optional: Calculate the imaginary roots.

You have now written a simple program to add two numbers on the DSP board and change the value of a variable using GUI and this ends Lab1.

2

Memory Organization and I/O Ports

2.1 Introduction

In the previous chapter we examined the following items:

- Difference between DSP, DSC, μP, μC and PC
- Peripherals
- Additional HW units of a DSP
- Real time control
- Demo board equipped with TI 320F28335 DSP
- Code Composer Studio
- General extension language (slider).

Also, the overview of Lab1 is as follows:

- Why were the source files were linked to the project?
- Why was the (float) written before the slider variable?

Remarks:

- C programming language is case sensitive!
- The local variables have to be declared right after the declaration of the function.
- Only one main per project.

In this section we will examine the memory organization and input/output ports. The structure of DSP 320F28335 is shown in Figure 2.1.

Figure 2.1: The structure of DSP 320F28335.

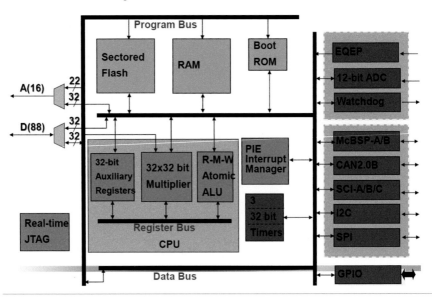

2.2 Memory Organization

The data storage for microprocessors include:

- Registers

    - Type: Volatile
    - Fastest access in 1 instruction cycle for read and write
    - Small in size
    - Cannot be used as program memory
    - Data stored is only temporary.

- Ram Random Access (RAM) memory

    - Type: Volatile
    - Fast access for read and write
    - Can work as program and also as data storage.

- Flash

o   Type: Nonvolatile
o   Usually slower than RAM for read
o   Write process is difficult, takes several clock cycles, it can be done for a limited number of times.

- External memory

o   Type: Volatile or nonvolatile
o   This memory communicates through the I/O ports of the µC
o   Slow access for read and write.

How does a memory work? The operation of memory is illustrated in the following block diagram of Figure 2.2.

Figure 2.2: The operation of memory.

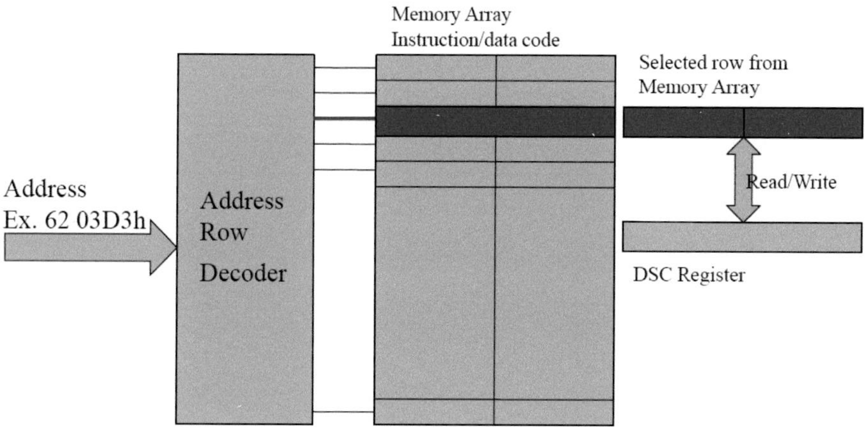

The important memory regions are as follows:

- 000 000 – 000 040 – Allocated start up address
- 000 D00 – 000 E00 – Interrupt vector table
- 008 000 – 00F 000 – RAM
- 330 000 – 33F FF8 – FLASH (ROM)
- 3FE 000 – 3FF FC0 – BOOT ROM

The conversion of the C code into machine code is indicated inside the red oval shown in the block diagram of Figure 2.3.

Figure 2.3: Conversion of the C code into machine code.

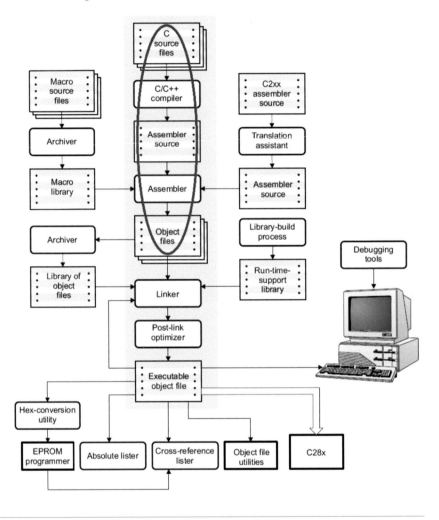

The C-compiler has the following features:

- Each "C" or "C++" code line is transformed into assembly language source code.
- The translated code can be stored .lib files.

    o Library is a collection of subroutines or classes used to develop software

The assembler-object file has the following features:

- .text usually contains executable code
- .data usually contains initialized data
- .bss usually reserves space for uninitialized variables.

For example, in the code below:

```
long z;
void main(void){
int x = 2;
int y = 7;
z = x + y;
while(1) asm("NOP");
}
```

the asm("NOP") is the .text type, int x = 2 is the .data type and long z is the .bss type. Therefore, the assembler transforms the source files into machine language object modules (.obj).

The linker combines object files into a single executable file where the segments are linked to memory addresses (.cmd), as illustrated in Figure 2.4.

Figure 2.4: The linker.

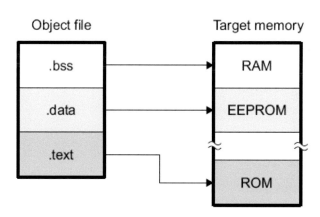

For example, the 28335_RAM_lnk.cmd code structure is as follows:

```
MEMORY{

PAGE 0 :

BEGIN : origin = 0x000000, length = 0x000002 //Boot

BOOT_RSVD : origin = 0x000002, length = 0x00004E

RAMM0 : origin = 0x000050, length = 0x0003B0

RAML0 : origin = 0x008000, length = 0x001000

RAML1 : origin = 0x009000, length = 0x001200

....

PAGE 1 :

RAMM1 : origin = 0x000400, length = 0x000400

RAML4 : origin = 0x00C000, length = 0x001000

RAML5 : origin = 0x00D000, length = 0x001000

....

}

SECTIONS{

codestart: > BEGIN, PAGE = 0

ramfuncs: > RAML0, PAGE = 0

.text : > RAML1, PAGE = 0

.cinit: > RAML0, PAGE = 0

.pinit: > RAML0, PAGE = 0

.switch : > RAML0, PAGE = 0

n.stack : > RAMM1, PAGE = 1

.ebss: > RAML4, PAGE = 1

.econst: > RAML5, PAGE = 1

.esysmem: > RAMM1, PAGE = 1

IQmath: > RAML1, PAGE = 0

}
```

The file DSP2833x_Headers_nonBIOS.cmd features are as follows:

- This file place the peripheral structures on the memory map locations.
- BIOS is an interface between to the hardware modules.
- Example:

GPIOCTRL : origin = 0x006F80, length = 0x000040 //ctrl. Reg.

GPIODAT : origin = 0x006FC0, length = 0x000020 //data reg.

GPIOINT : origin = 0x006FE0, length = 0x000020 //Interrupt

The stack application includes the following features:

- Ram memory section used for temporary storage of the processor status
- Example:

long z, x;

void calc_sum(intx , inty){

z = x + y;

}

void main(void){

x = 3;

calc_sum(2,7); //Save the register status in stack

x += z; //Reload the register status from stack

while(1) asm(" NOP");

}

2.3 I/O Ports

The I/O ports have the following features:

- Through the I/O ports a communication is made between the μP and the outside world.
- Typical output port construction:

- o   Open collector
- o   Transistor-transistor logic.

- Typical input port construction:

    - o   Transistor–transistor logic
    - o   Schmitt trigger.

The output port characteristics are as follows:

- Simplified schematic of an open collector circuit is shown in Figure 2.5.

    - o   Independent of the DSP voltage
    - o   Easy to use for communication
    - o   High energy consumption.

Figure 2.5: Simplified schematic of an open collector circuit.

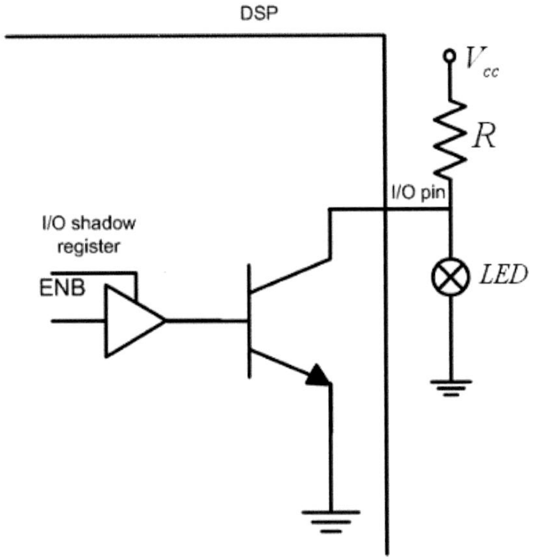

- A simplified schematic of transistor–transistor logic is depicted in Figure 2.6.

    - o   No external circuit is needed to set the output voltage
    - o   Difficult to use for communication
    - o   Low energy consumption.

Figure 2.6: Simplified schematic of transistor–transistor logic.

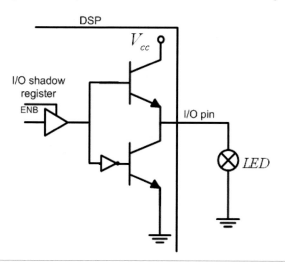

The input port characteristics are as follows:

- Transistor–transistor logic is illustrated in Figure 2.7:

  - The logic 1 state has to be higher than $V_{cc}/2$
  - Usually, it can tolerate around 15% higher voltage than V_{cc}.

Figure 2.7: The Transistor-Transistor logic.

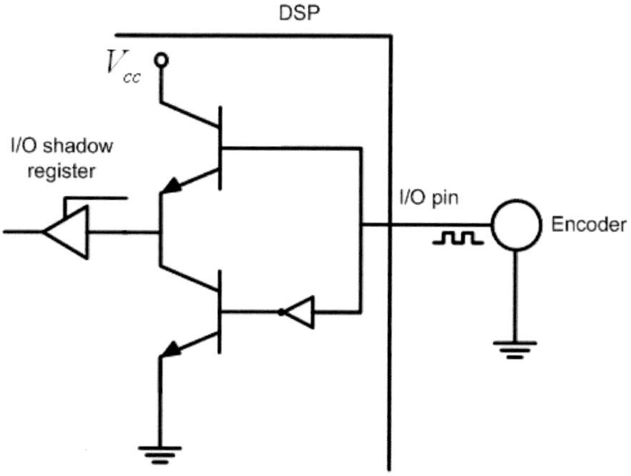

- Schmitt trigger:

  ○ Usually used for oscillator input, external reset, and external interrupt, as illustrated in Figure 2.8.
  ○ Has a hysteresis on the input voltage; for ex. at 5 V operation above 3.5 V is logical "1", below 1.5 V is logical "0" (with 0–3.3 V interface will not work).

Figure 2.8: The Schmitt trigger.

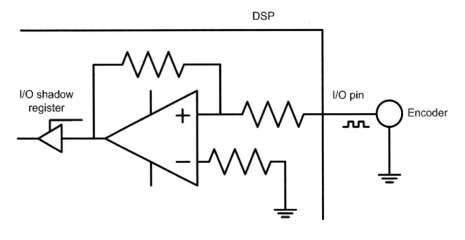

The I/O registers of 320F28335 are as follows:

- Two structures are attached to I/O registers:

  ○ GpioCtrlRegs
  ○ GpioDataRegs

- GpioCtrlRegs.GPAPUD.bit.GPIO17

  ○ 0 Enable pull up resistor
  ○ 1 Disable pull up resistor

- GpioCtrlRegs.GPAMUX2.bit.GPIO17

  ○ 0 the port is controlled by the microprocessor
  ○ 1 the port is controlled by the peripheral

- GpioCtrlRegs.GPADIR.bit.GPIO17

  - ○ 1 the port is output
  - ○ 0 the port is input

- GpioDataRegs.GPASET.bit.GPIO18 = 1

  - ○ the output port will be high

- GpioDataRegs.GPACLEAR.bit.GPIO18 = 1

  - ○ the output port will be low

- GpioDataRegs.GPADAT.bit.GPIO18 = x

  - ○ the output port will have the value of x (can be 1 or 0)

- GpioDataRegs.GPATOGGLE.bit.GPIO18

  - ○ the output port will change its value (from 1 to 0 or from 0 to 1)

- Certain memory ranges are protected by EALLOW-EDIS.

Therefore, we can summarize this section as follows:

- Memory organization

  - ○ Types of memory
  - ○ Memory structure of 320F28335

- Digital input–output port

  - ○ Different hardware construction
  - ○ I/O ports of 320F28335.

2.4 Lab2

In this section, we want to toggle two LEDs connected to GPIO0 and GPIO2 pins with a delay defined inside the dealy_loop() function; you can see the lab2.c code contents in Figure 2.9.

Figure 2.9: The lab2.c code contents.

```
 1 /*
 2  * lab2.c
 3  *
 4  *
 5  *
 6  *
 7  */
 8 #include "IO.h"
 9 #include "DSP2833x_Device.h"      // DSP2833x Header file Include File
10 #include "DSP2833x_Examples.h"
11
12 void delay_loop ();
13
14 void main(void){
15     InitSysCtrl();
16     DINT;
17     InitPieCtrl();
18     IER = 0x0000;
19     IFR = 0x0000;
20     InitPieVectTable();
21
22     IO_Init();
23
24     while(1){
25         IO0_Toggle();
26         delay_loop();
27         IO2_Toggle();
28         delay_loop();
29     }
30 }
31 void delay_loop()
32 {
33     unsigned int     i;
34     for (i = 0; i <300; i++);
35 }
```

Also, the code contents of the files IO.c and IO.h are illustrated in Figure 2.10 and Figure 2.11 respectively. Moreover, after building and running the code, connect the oscilloscope to I/O number 0 and 2 (GPIO0 and GPIO2). You should see two rectangular signals on the screen of the oscilloscope as shown in Figure 2.12.

In Lab2 we will do the following tasks:

- Task 1: modify the code on a way to be able to change the frequency of the signal with a slider.
- Task 2: Modify the code to be able to change the frequency of one channel compared to the other one by using another slider.

Figure 2.10: The IO.c code contents.

```
 Lab1.c    NewTargetConfiguration.ccxml     IO.h    IO.c ⊠    lab2.c
 1 /*
 2  * IO.c
 3  *
 4  *
 5  *
 6  *
 7  */
 8
 9 #include "IO.h"
10
11 void IO_Init(){
12
13 EALLOW;
14     GpioCtrlRegs.GPAPUD.bit.GPIO0    = 0; // Enable pullup on GPIO0
15     GpioCtrlRegs.GPAMUX1.bit.GPIO0   = 0; // GPIO0 = I/O
16     GpioCtrlRegs.GPADIR.bit.GPIO0    = 1; // GPIO0 = output
17
18     GpioCtrlRegs.GPAPUD.bit.GPIO2    = 0; // Enable pullup on GPIO2
19     GpioCtrlRegs.GPAMUX1.bit.GPIO2   = 0; // GPIO2 = I/O
20     GpioCtrlRegs.GPADIR.bit.GPIO2    = 1; // GPIO2 = output
21
22     GpioCtrlRegs.GPAPUD.bit.GPIO4    = 0; // Enable pullup on GPIO4
23     GpioCtrlRegs.GPAMUX1.bit.GPIO4   = 0; // GPIO4 = I/O
24     GpioCtrlRegs.GPADIR.bit.GPIO4    = 1; // GPIO4 = output
25 EDIS;
26 }
```

Figure 2.11: The IO.h code contents.

```
 pwm.c    pwm.h    main.c    main.c    lab2.c    IO.h ⊠
 1 /*
 2  * IO.h
 3  *
 4  *
 5  *
 6  */
 7
 8 #ifndef IO_H_
 9 #define IO_H_
10
11 #include "DSP2833x_Device.h"      // DSP2833x Headerfile Include File
12 #include "DSP2833x_Examples.h"
13
14 #define IO0_ON()        GpioDataRegs.GPASET.bit.GPIO0 = 1
15 #define IO0_OFF()       GpioDataRegs.GPACLEAR.bit.GPIO0  = 1
16 #define IO0(x)          GpioDataRegs.GPADAT.bit.GPIO0  = x
17 #define IO0_Toggle()    GpioDataRegs.GPATOGGLE.bit.GPIO0 = 1
18 #define IO2_ON()        GpioDataRegs.GPASET.bit.GPIO2 = 1
19 #define IO2_OFF()       GpioDataRegs.GPACLEAR.bit.GPIO2  = 1
20 #define IO2(x)          GpioDataRegs.GPADAT.bit.GPIO2  = x
21 #define IO2_Toggle()    GpioDataRegs.GPATOGGLE.bit.GPIO2 = 1
22 #define IO4_OFF()       GpioDataRegs.GPACLEAR.bit.GPIO4  = 1
23
24 void IO_Init();
25 #endif /* IO_H_ */
```

Figure 2.12: The generated pulses at the output pines, GPIO0 and GPIO2.

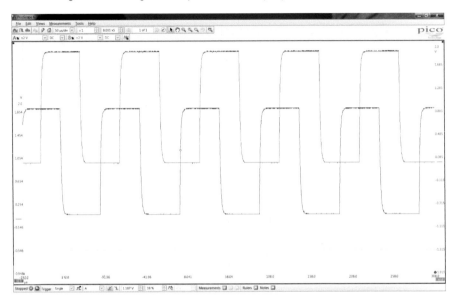

In this lab, you will create a rectangular signal with variable frequency using an output port of the DSC. For generation of a variable frequency rectangular signal do the following steps:

(1) Exclude Lab1.c file from the project directory (right click on the file name and select exclude from project).

(2) Create a new source file (File/New/Source File) with the name of lab2.c, IO.c, and a new header file (File/New/Source File) IO.h as in Figure 2.13.

(3) Insert the next code in the IO.c file:

#include "IO.h"

void IO_Init()

{

EALLOW;

GpioCtrlRegs.GPAPUD.bit.GPIO0 = 0; // Enable pullup on GPIO0

GpioCtrlRegs.GPAMUX1.bit.GPIO0 = 0; // GPIO0 = I/O GpioCtrlRegs.GPADIR. bit.GPIO0 = 1; // GPIO0 = output

Figure 2.13: Creating new source and header files.

GpioCtrlRegs.GPAPUD.bit.GPIO2 = 0; // Enable pullup on GPIO2

GpioCtrlRegs.GPAMUX1.bit.GPIO2 = 0; // GPIO2 = I/O GpioCtrlRegs.GPADIR. bit.GPIO2 = 1; // GPIO2 = output

GpioCtrlRegs.GPAPUD.bit.GPIO4 = 0; // Enable pullup on GPIO4

GpioCtrlRegs.GPAMUX1.bit.GPIO4 = 0; // GPIO4 = I/O GpioCtrlRegs.GPADIR. bit.GPIO4 = 1; // GPIO4 = output

EDIS;

}

(4)   Insert the next code in the IO.h file:

#include "DSP2833x_Device.h" // DSP2833x Headerfile Include File

#include "DSP2833x_Examples.h"

#define IO0_ON() GpioDataRegs.GPASET.bit.GPIO0 = 1

#define IO0_OFF() GpioDataRegs.GPACLEAR.bit.GPIO0 = 1

```
#define IO0(x) GpioDataRegs.GPADAT.bit.GPIO0 = x
#define IO0_Toggle() GpioDataRegs.GPATOGGLE.bit.GPIO0 = 1
#define IO2_ON() GpioDataRegs.GPASET.bit.GPIO2 = 1
#define IO2_OFF() GpioDataRegs.GPACLEAR.bit.GPIO2 = 1
#define IO2(x) GpioDataRegs.GPADAT.bit.GPIO2 = x
#define IO2_Toggle() GpioDataRegs.GPATOGGLE.bit.GPIO2 = 1
void IO_Init();
```

(5)   Insert the next code in the lab2.c file:

```
#include "IO.h"
#include "DSP2833x_Device.h" // DSP2833x Header file Include File
#include "DSP2833x_Examples.h"
void delay_loop ();
void main(void){
InitSysCtrl();
DINT;
InitPieCtrl();
IER = 0x0000;
IFR = 0x0000;
InitPieVectTable();
IO_Init();
while(1){
IO0_Toggle();
delay_loop();
IO2_Toggle();
delay_loop();
}
}
```

```
void delay_loop()

{

unsigned int i;

for (i = 0; i <300; i++);

}
```

(6)   Build and run the code.

(7)   Connect the oscilloscope to I/O number 0 and 2.

(8)   You should see two rectangular signals on the screen of the oscilloscope as depicted in Figure 2.12.

(9)   Your first task is to modify the code on a way to be able to change the frequency of these two rectangular signals by using the slider from the previous Lab1.

(10) The second task is to change the frequency of the first channel signal (blue signal) compared to the frequency of the second channel signal as in Figure 2.14.

Figure 2.14: Changing the frequency of the first channel signal compared to the frequency of the second channel signal.

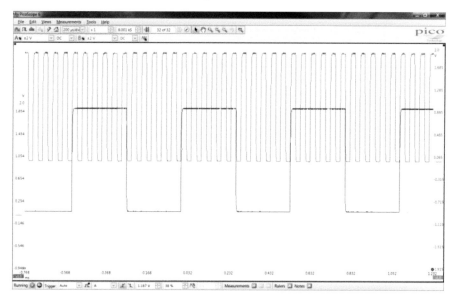

(11)   Create a new slider.

(12)   Modify the code to be able to change the frequency of one channel compared to the other with the second slider.

Task1 solution:

For using a slider do the following steps:

(1)   Create a new file using File\New\File with the name Input_Val1.gel

(2)   Place the next cod in the file:

menuitem "Input_Val1"

slider Input_Val1 (0,50,1 ,1, Input_Val1){

Input1= Input_Val1;

}

(3)   Save the file as slider.gel in the Lab project directory.

(4)   Correct the lab2.c code as follows:

```
#include "IO.h"

#include "DSP2833x_Device.h" // DSP2833x Header file Include File

#include "DSP2833x_Examples.h"

void delay_loop();

void main(void){

InitSysCtrl();

DINT;

InitPieCtrl();

IER = 0x0000;

IFR = 0x0000;

InitPieVectTable();

IO_Init();

while(1){

IO0_Toggle();

delay_loop();
```

```
IO2_Toggle();

delay_loop();

}

}

void delay_loop()

{

unsigned int i;

for (i = 0; i < Input1; i++);

}
```

Task2 solution:

Correct the lab2.c code as follows.

```
#include "IO.h"

#include "DSP2833x_Device.h" // DSP2833x Header file Include File

#include "DSP2833x_Examples.h"

void delay_loop1();

void main(void){

int k = 0;

InitSysCtrl();

DINT;

InitPieCtrl();

IER = 0x0000;

IFR = 0x0000;

InitPieVectTable();

IO_Init();

while(1){

IO0_Toggle();

delay_loop1();

k = k + 1;
```

```
if (k == Input1) {

k = 0;

IO2_Toggle();

}

}

}

void delay_loop1()

{

unsigned int j;

for (j = 0; j < 10; j++);

}
```

Task3 solution:

For using a slider do the following steps:

(1) Create a new file using File\New\File with the name Input_Val2.gel
(2) Place the next cod in the file:

```
menuitem "Input_Val2"

slider Input_Val2 (0,50,1 ,1, Input_Val2){

Input2= Input_Val2;

}
```

(3) Save the file as slider.gel in the Lab project directory.
(4) Correct the lab2.c code as follows:

```
#include "IO.h"

#include "DSP2833x_Device.h" // DSP2833x Header file Include File

#include "DSP2833x_Examples.h"

void delay_loop1 ();

void main(void){

int k=0;
```

```
InitSysCtrl();

DINT;

InitPieCtrl();

IER = 0x0000;

IFR = 0x0000;

InitPieVectTable();

IO_Init();

while(1){

IO0_Toggle();

delay_loop1();

k = k + 1;

if (k == Input1) {

k = 0;

IO2_Toggle();

}

}

}

void delay_loop1()

n{

unsigned int j;

for (j = 0; j < Input2; j++);

}
```

The other solution for Lab2 is the following code, where we use an int function instead of a void function.

```
#include"IO.h"

#include"DSP2833x_Device.h"// DSP2833x Header file Include File

#include"DSP2833x_Examples.h"
```

```
void delay_loop_1 (int freq);

intfreq1,freq2;

inti;

void main(void){

InitSysCtrl();

DINT;

InitPieCtrl();

IER = 0x0000;

IFR = 0x0000;

InitPieVectTable();

IO_Init();

while(1){

for(i=0;i<freq2;i++){

IO2_Toggle();

delay_loop_1(freq1);

}

IO0_Toggle();

IO2_Toggle();

delay_loop_1(freq1);

}

}

void delay_loop_1 (int freq)

{

Unsigned int i;

for(i = 0; i <freq; i++);

}
```

3

System Interrupts and Timers

3.1 Introduction

In the previous chapter we examined the following items:

- Memory organization

  - Types of memories
  - How a C code is transformed into assembly code, and the way this code is placed into the physical memory of the DSP.

- Input–output ports

  - Hardware configuration
  - Control registers for the I/O ports.

Also, we did the following tasks in the Lab2 section:

- Generation of a variable frequency rectangular signal by using I/O ports of the DSC.
- For each I/O port has to be set:

  - The internal pull-up resistors have to be enabled
  - The direction of the port has to be set
  - The I/O pin has to be controlled by the DSC.

- First task: change the frequency of the rectangular signal by using a slider.
- Change the frequency of the first signal compared to the second signal.

Furthermore, as reminder for C programming, we can answer the following questions:

- Is C programing case sensitive?
- Why do we use functions?
- What is the difference between a void and an int function?
- Why do we declare the header of the function before the main and have the instruction written after the main?

The agenda of this chapter is to examine the following items:

- System interrupts
- Timers.

3.2 System Interrupts

The system interrupts characteristics are as follows:

- Interrupt stops the running program to start a higher important task called the "interrupt subroutine".
- Most of the peripheral modules (ADC, PWM, Timer etc.) can generate interrupts.
- Advantage of the usage of interrupts is:

  o The ALU does not have to wait until the HW module is working
  o Gives fast reaction on environmental changes (over current, over temperature etc.).

The interrupt mechanism is shown in Figure 3.1.

Figure 3.1: The interrupt mechanism.

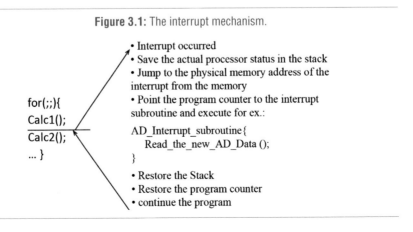

```
for(;;){
Calc1();
Calc2();
... }
```

- Interrupt occurred
- Save the actual processor status in the stack
- Jump to the physical memory address of the interrupt from the memory
- Point the program counter to the interrupt subroutine and execute for ex.:

AD_Interrupt_subroutine{
    Read_the_new_AD_Data ();
}

- Restore the Stack
- Restore the program counter
- continue the program

There are two types of interrupts which are:

- Maskable interrupts, which can be enabled or disabled from software

  o Usually, one bit is set by the HW signalizing the interrupt request, another bit(s) has to be set by the user which can allow or forbid the interruption of the CPU process.

- Non-maskable interrupts, which cannot be disabled

  o The actual CPU process is interrupted (highest priority task).

The system interrupts for 320f28335 are depicted in Figure 3.2.

Figure 3.2: The system interrupts for 320f28335.

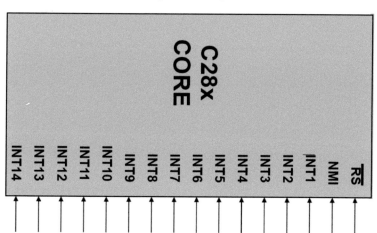

- 2 non-maskable interrupts (RS, "selectable" NMI)
- 14 maskable interrupts (INT1–INT14).

The maskable interrupt processing block diagram is illustrated in Figure 3.3.

The maskable interrupt processing includes the following characteristics:

- A valid signal on a specific interrupt line causes the latch to display a "1" in the appropriate bit.
- If the individual and global switches are turned "on" the interrupt reaches the core.

Figure 3.3: The maskable interrupt processing.

Core Interrupt	(IFR) "Latch"	(IER) "Switch"	(INTM) "Global Switch"	
$\overline{\text{INT1}}$				
$\overline{\text{INT2}}$				C28x Core
$\overline{\text{INT14}}$				

An overview of peripheral interrupt expansion (PIE) is shown in Figure 3.4.

Figure 3.4: The overview of peripheral interrupt expansion (PIE).

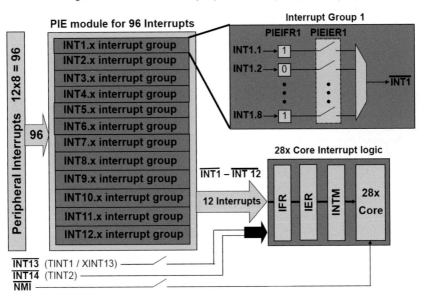

The operation of two interrupt files, PieCtrl.c and PieVect.c are as follows:

- Void InitPieCtrl(void) – disables the CPU level interrupts
- Void EnableInterrupts(void) – enables the peripheral interrupts
- InitPieVectTable(void) – initializes the interrupt vector table.

An example of the C28x timer interrupt (TINT) system is illustrated in Figure 3.5.

Figure 3.5: An example of C28x timer interrupt (TINT) system.

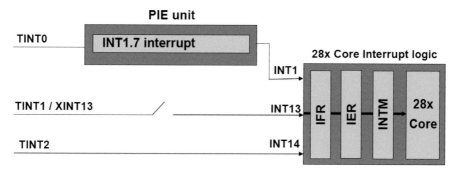

The interrupt response of hardware sequence is shown in Figure 3.6.

Figure 3.6: The interrupt response of hardware sequence.

CPU Action	Description
Registers → stack	14 Register words auto saved
0 → IFR (bit)	Clear corresponding IFR bit
0 → IER (bit)	Clear corresponding IER bit
1 → INTM/DBGM	Disable global ints/debug events
Vector → PC	Loads PC with int vector address
Clear other status bits	Clear LOOP, EALLOW, IDLESTAT

Note that some actions occur simultaneously, and none are interruptible.

A summary of interrupts is:

- Interrupt stops the running program to start a higher important task (interrupt subroutine).
- Two types of interrupts: maskable and non-maskable.
- 96 PIE interrupts can be generated in 320F28335.

3.3 Timers

The timers' characteristics are as follows:

- Timers are digital counters which have an increment or decrement at a fixed frequency (equal or less than the µC clock).
- Main characteristics: Time base (TB) and period

  o Period register (PR) contains an integer number which shows when the counter has to be restarted.
  o The TB usually is synchronized with the system clock.

The waveform of the timer main characteristics is shown in Figure 3.7.

Figure 3.7: The waveform of timer main characteristics.

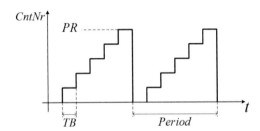

The two types of counters are depicted in Figure 3.8.

Figure 3.8: The two types of counters.

$$TB = \frac{n}{\mu C\_Clock\_freq}$$ n – is the pre-scale system clock divider

Down counter   Up-Down counter

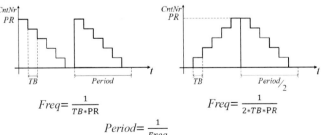

$$Freq = \frac{1}{TB*PR}$$ $$Freq = \frac{1}{2*TB*PR}$$

$$Period = \frac{1}{Freq}$$

In 320F28335, there are three 32bit timers, as illustrated in Figure 3.9.

Figure 3.9: The three timers' structure in 320F28335.

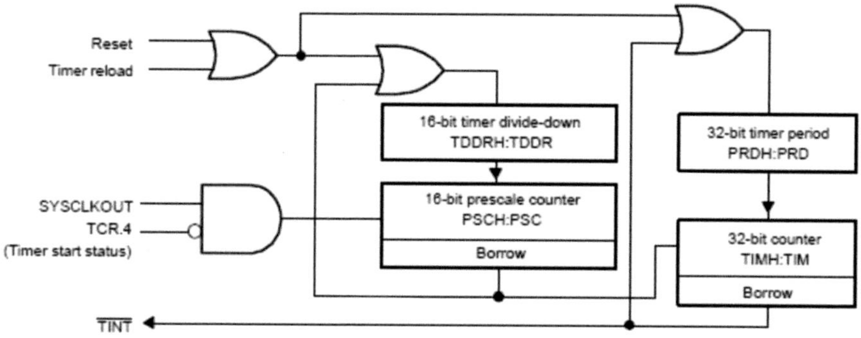

The timer initialization codes in C are as follows:

void InitTimer(void){

CpuTimer0Regs.PRD.all = 0xFFFFFFFF; //Initialize the PR register

CpuTimer0Regs.TPR.all = 0; //Initialize pre-scale counter

CpuTimer0Regs.TPRH.all = 0;

CpuTimer0Regs.TCR.bit.TSS = 1; // Make sure timer is stopped after //initialization

CpuTimer0Regs.TCR.bit.TRB = 1; // Reload all counter register with period //value

CpuTimer0Regs.TCR.bit.SOFT = 0; //These bits are special emulation bits

CpuTimer0Regs.TCR.bit.FREE = 0;

}

The timer interrupt in C is implemented as follows:

- Declaration of an interrupt function:
  interrupt void cpu_timer0_isr(void);
- Association of the interrupt function with the interrupt vector table

EALLOW;

*(Uint32 *) &PieVectTable.TINT0 = (Uint32)cpu_timer0_isr;

EDIS;

The code for creating a watch is as follows:

```
void Timing_Run(Type_TIMING*BaseTime){

BaseTime->uSec+= TimerTimeBase;

if(BaseTime->uSec>= 1000){

BaseTime->uSec= 0;

BaseTime->mSec+= 1;

}

if(BaseTime->mSec>= 1000){

BaseTime->mSec= 0;

BaseTime->Sec += 1;

YourSecTiming++;

}

}//void Timing_Run

typedefstruct{ unsigned intuSec;

unsigned intmSec;

unsigned intSec;

} Type_TIMING;

main(){ ...

while(1){

if(YourSecTiming>10){

YourSecTiming=0;

>run the code with

one second refresh<

}
```

}

For parallel calculation using timer (multithreading) we should do as follows:

- Each timer interrupt can be set to handle an independent process in parallel with the process from the main loop
- Example:

//elevator

while(1){

if(button_pressed){

identify_button();

close_door();

move_elevator();

open_door();

...}

}

//temperature control inside the cabin

interrupt void cpu_timer0_isr(){

//interrupt in each second

if(temperature>ref)

   Heating_off();

else

   Heating_on();

The summary of timers are as follows:

- Timers are digital counters which have an increment or decrement at a fixed frequency.
- Timers can be set to generate an interrupt.
- Timers can be used as an additional thread to the main loop.

3.4 Lab3

We will examine the following exercises in this section:

- Task 1: modify the given code in a way that the input parameter for the set function for the timer (SetTimer) to be in frequency (Hz) or in period of time (µs)!
- Task 2: Modify the code to be able to change the frequency of the two channels independently by using two sliders.
- Task 3: Create a rectangular signal like that in Figure 3.10.

Figure 3.10: The rectangular signal.

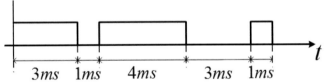

In this lab, you will create a rectangular signal with variable frequency by using a timer interrupt. For generating of a variable frequency rectangular signal by using a timer do the following steps:

(1) Exclude the Lab2.c from the project.
(2) Add the new source files: Lab3.c, Timer.c and Timer.h to the project.

The source files, Lab3.c, Timer.c and Timer.h are as follows:

Lab3.c:

```
/*

* Lab3.c

*

*

*/
```

#include "IO.h"

#include "DSP2833x_Device.h" // DSP2833x Headerfile Include File

#include "DSP2833x_Examples.h"

```
#include "Timer.h"
interrupt void cpu_timer0_isr(void);
// Global Variables
int Input=2, freq=50;
int i=0;
void main(void){
   InitSysCtrl();
   DINT;//disable interrupts
   InitPieCtrl();
   IER = 0x0000; //disable all interupts
   IFR = 0x0000; //clear all interupts flags
   InitPieVectTable();
   // Interrupts that are used in this example are //re-mapped to
   EALLOW; // This is needed to write to EALLOW protected register
   *(Uint32 *) &PieVectTable.TINT0 = (Uint32)cpu_timer0_isr;
   //The cpu_timer0_isr function is associated to //the timer interrupt
   EDIS; // This is needed to disable write to EALLOW protected registers
   DINT;//disable interrupts
   InitTimer();
   IO_Init();
   SetTimer(30.0);//time in micro seconds
   IER = M_INT1|M_INT3;
   // Enable TINT0 in the PIE: Group 1 interrupt 7
   PieCtrlRegs.PIEIER1.bit.INTx7 = 1;
   EINT; // Enable Global interrupt INTM
   ERTM; // Enable Global realtime interrupt DBGM
   StartCpuTimer0();
```

```
    while(1){
    } //while(1){ infinite waiting loop
}

interrupt void cpu_timer0_isr(void){
        IO0_Toggle();
        if(i++>Input){
        IO2_Toggle();
        i = 0;
        }
        SetTimer(((float)freq));
        // Acknowledge this interrupt to receive more //interrupts from group 1
        PieCtrlRegs.PIEACK.all = PIEACK_GROUP1;
}
```

Timer.c:

```
/*
* Timer.c
*/
#include "DSP2833x_Device.h" // Headerfile Include File
#include "DSP2833x_Examples.h" // Examples Include File
#include "Timer.h"

//This function will initialize CPU Timer 0
void InitTimer(void){
    // Initialize timer period to maximum:
    CpuTimer0Regs.PRD.all = 0xFFFFFFFF;// Initialize //pre-scale counter to
divide by 1 (SYSCLKOUT):
    CpuTimer0Regs.TPR.all = 0;
```

CpuTimer0Regs.TPRH.all = 0;// Make sure timer is //stopped after initialization:

CpuTimer0Regs.TCR.bit.TSS = 1;// Reload all //counter register with period value:

CpuTimer0Regs.TCR.bit.TRB = 1;

//These bits are special emulation bits that //determine the

//SOFT state of the timer when a breakpoint is //encountered in the high-level language

CpuTimer0Regs.TCR.bit.SOFT = 0;

CpuTimer0Regs.TCR.bit.FREE = 0;

}

// ConfigCpuTimer:

void SetTimer(float Period_us){

CpuTimer0Regs.PRD.all = (Uint32)(Period_us);

// Initialize timer control register:

//reload timer

CpuTimer0Regs.TCR.bit.TRB = 1;

//interrupt enable

CpuTimer0Regs.TCR.bit.TIE = 1;

StartCpuTimer0();

}

//==

// End of file.

//==

Timer.h:

/*

* Timer.h

*/

```
#ifndef TIMER_H_

#define TIMER_H_

#define CPU_Speed 150e6

#define CPU_Speed_us 150.0

void SetTimer(float Period);

void InitTimer(void);

#endif /* TIMER_H_ */
```

(3)   Build and run the code.

(4)   Connect the oscilloscope to pin number 0 and 02.

(5)   You should see two rectangular signals on the screen of the oscilloscope like during the previous lab, as shown in Figure 3.11.

Figure 3.11: Two rectangular signals on the screen of oscilloscope.

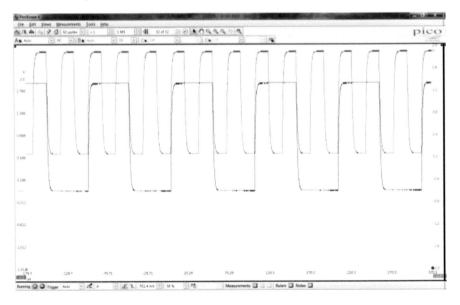

(6)   Notice, now the I/O pins are controlled inside the timer interrupt; the infinite loop from the main function could execute a parallel thread.

(7)   Your first task is to modify the given code in a way, that the input parameter for the set function (SetTimer) to be in frequency (Hz) or in period of time (μs).

(8)   Example: when the slider shows 200 (μs), you should have on the scope also 200 μs, as in Figure 3.12.

Figure 3.12: Square wave with period of 200μs.

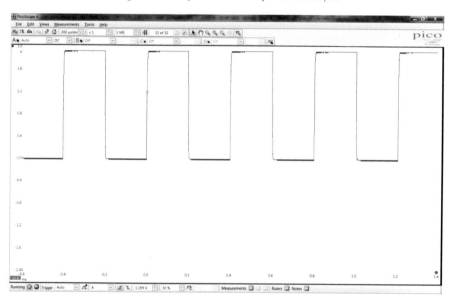

The first task is to modify the given code on a way that the input parameter for the SetTimer function to be in frequency (Hz) or in period of time (s or μs). A solution could be:

#define CPU_Speed (float)150.0 //in MHz

void SetTimer(float Period){//Period is in us

PR = (Uint32) (Period * CPU_Speed);

The square wave on oscilloscope screen is depicted in Figure 3.13.

(9)   Second task is to modify the code to be able to change the frequency of the two channels independently by using two sliders, as illustrated in Figure 3.14.

(10)   As can be seen in Figure 3.14, the two signals are totally independent.

Figure 3.13: The square wave on oscilloscope screen.

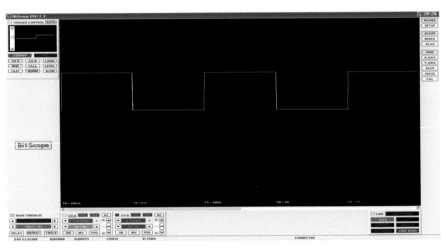

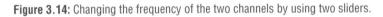

Figure 3.14: Changing the frequency of the two channels by using two sliders.

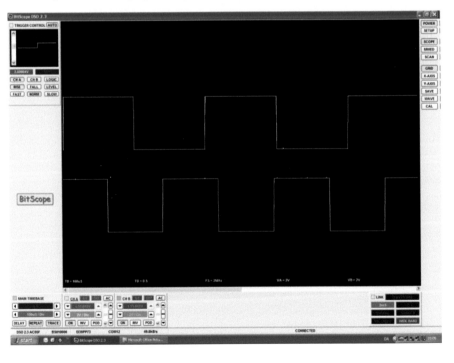

The second task is to modify the code to be able to change the frequency of the two channels independently.

A solution would look like:

```
while(1){

IO0_Toggle();

delay_loop1(Per1);

}

void delay_loop1 (int freq)

{

unsigned int i;

for(i = 0; i < freq; i++);

}

slider1.gel:

menuitem "slider1"

slider slider1(0,1500,1,1,slider1){

Per1 = slider1;

}
```

Also, we do the following correction:

```
interrupt void cpu_timer0_isr(void){

    if(i++>Input){

    IO2_Toggle();

    i = 0;

    }

    SetTimer(((float)freq));

    // acknowledge this interrupt to //receive more //interrupts from group 1

    PieCtrlRegs.PIEACK.all = PIEACK_GROUP1;

}
```

slider2.gel:

menuitem "slider2"

slider slider2(0,1500,1,1,slider2){

Input = slider2;

}

(11)   Modify the code in order to create a train of pulses with the duration 1 ms; 3 ms; 1 ms; 4 ms; 3 ms, as in Figure 3.15.

Figure 3.15: Creating a train of pulses.

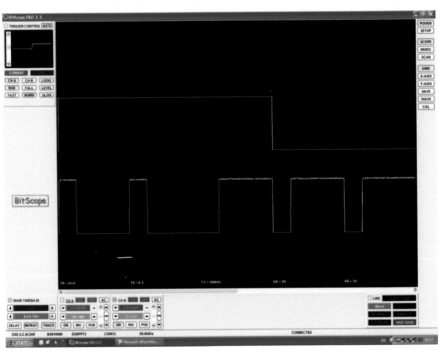

The third task is to modify the code to create a train of pulses with the duration of 1 ms; 3 ms; 1 ms; 4 ms; 3 ms.

A solution would look like:

int Scheme[5]={1,3,1,4,3}, index= 0;

```
interrupt void cpu_timer0_isr(void){

IO18_Toggle();

SetTimer(((float)Scheme[index++])*1e3);

if(index > 5) index = 0;

PieCtrlRegs.PIEACK.all = PIEACK_GROUP1;

}
```

4

Digital to Analog Conversion

4.1 Introduction

In the previous chapter we examined the following items:

- System interrupts

    - Interrupt stops the running program to start a higher important task (the interrupt subroutine).
    - Two types of interrupts: maskable and non-maskable.
    - 96 PIE interrupts can be generated in 320f28335 (most of the peripheral can generate an interrupt).

- Timer

    - Timers are digital counters that have an increment at a fixed frequency (equal or less than the μC clock); reaching the PR value it is reset to zero.
    - Timers can be set to generate an interrupt at reset.
    - Timers can be used as an additional thread to the main loop.

4.2 The Realization of Digital to Analog Conversion (DAC)

The agenda of this chapter is as follows:

Digital to analog conversion (DAC):

- What is DAC?
- Requirements of DAC.
- Theory behind PWM.
- PWM module of TMS320F28335.

What is a digital to analog converter?

- Converts a digital number to analog voltage.
- It can be considered as the inverse of ADC.

The block diagram of a digital to analog converter (DAC) is illustrated in Figure 4.1.

Figure 4.1: The block diagram of a digital to analog converter (DAC).

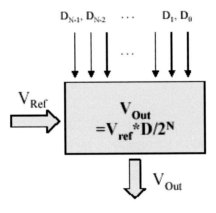

There are three main parameters that should be considered when choosing a DAC for a particular project:

- Resolution
- Reference voltage
- Time needed for one conversion.

The resolution characteristics are as follows:

- The resolution is the amount of voltage rise created by increasing the least significant bit (LSB) of the input by 1. This voltage value is a function of the number of input bits and the reference voltage value with the following equation:

$$Resolution = \frac{Reference\_Voltage}{2^{n\_bits}}. \tag{4.1}$$

- Increasing the number of bits results in a finer resolution.
  The reference voltage characteristics are as follows:
- To a large extent the output properties of a DAC are determined by the reference voltage (maximum output voltage).
- Multiplier DAC – the reference voltage is constant and is set by the manufacturer.
- Non-multiplier DAC – the reference voltage can be changed during operation.
  Also, the time needed for one conversion characteristics are as follows:
- Usually specified as the conversion rate or sampling rate. It is the rate at which the input register is cycled through in the DAC.
- High speed DACs are defined as operating at greater than 1 microsecond per sample (1 MHz).
- Some state of the art 12–16-bit DACs can reach speeds of 1 GHz.
- The conversion of the digital input signal is limited by the clock speed of the input signal and the settling time of the DAC as shown in Figure 4.2.

Figure 4.2: The conversion of the digital input signal limitations.

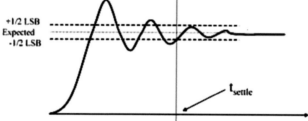

There are two different ways to implement DACs, which are as follows:

- Resistor string
- Digital potentiometer
- DAC circuit
- PWM.

The resistor string DAC has the following components, as depicted in Figure 4.3.

- Resistor string
- Selection switches
- Operational amplifier.

Figure 4.3: The resistor string DAC.

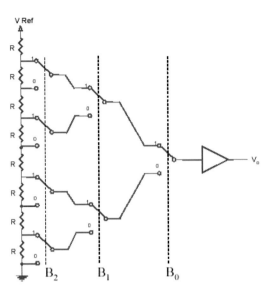

The digital potentiometer IC (MCP4011) structure is illustrated in Figure 4.4.

Figure 4.4: The digital potentiometer IC (MCP4011).

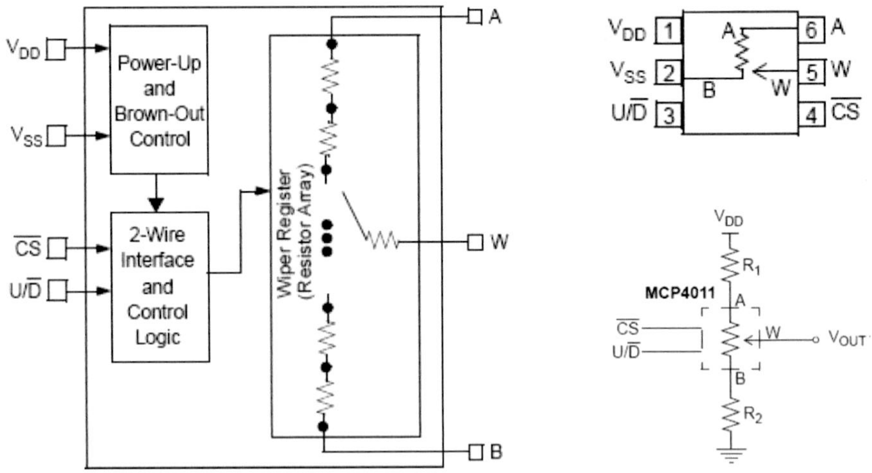

The typical DAC circuit IC (AD5724R-12 bit/AD5734R-14 bit/AD5754R-16 bit) structure is shown in Figure 4.5.

Figure 4.5: The typical DAC circuit IC.

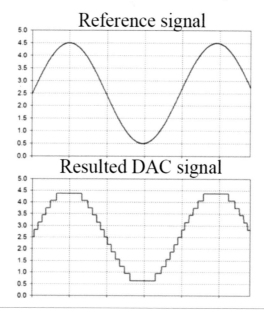

A comparison between the continuous signal and a DAC is depicted in Figure 4.6.

Figure 4.6: A comparison between the continuous signal and a DAC.

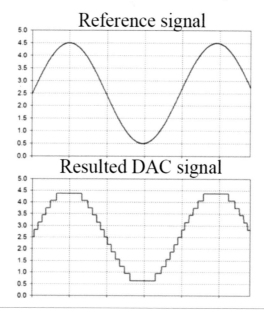

The pulse width modulation (PWM) has the following characteristics:

- Approximation of an analogue signal by switching in an on/off manner of a DC voltage.
- The average of the output voltage in a period is proportional to the reference voltage, as illustrated in Figure 4.7.

Figure 4.7: The PWM average is proportional to the reference voltage.

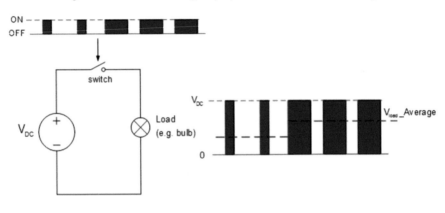

The typical application for PWM includes the following items:

- Converter applications (e.g., motor drives, etc.)
- Telecommunications
- Lamp dimmers
- Switching (audio) amplifiers
- Power delivery (SMPS, chargers, etc.)
- Voltage regulation
- Etc.

Also, a typical application for PWM is to use it for controlling AC motors and DC motors. The AC motor control example is shown in Figure 4.8. Furthermore, the DC motor control example using PWM is depicted in Figure 4.9.

The comparison between PWM versus resistive dividers are as follows:

- The DACs based on resistive dividers have the disadvantage of being applicable only for low power due to the high losses in the resistors.
- The PWM has a high efficiency (above 95% – depending on the switching device), so it is very often used in power electronics for electric motor control, energy transfer, etc.

Figure 4.8: The AC motor control example using PWM.

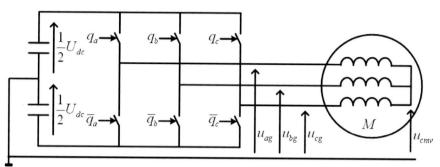

Figure 4.9: The DC motor control example using PWM.

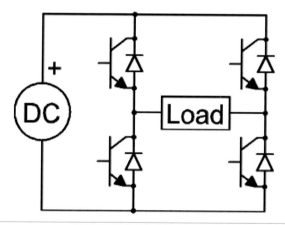

There are three ways to generate a pulse which include as below:

- Analog PWM
- Digital PWM with sawtooth carrier
- Digital PWM with triangular carrier.

The analog PWM has the following characteristics as illustrated in Figure 4.10.

- Reference signal > triangular carrier wave – ON
- Reference signal < triangular carrier wave – OFF.

Figure 4.10: The analog PWM.

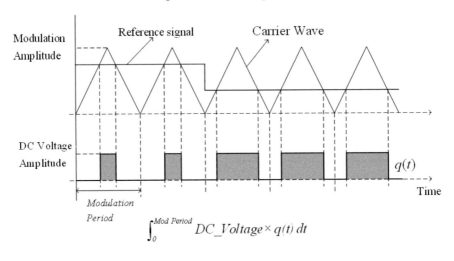

$$\int_{0}^{Mod\ Period} DC\_Voltage \times q(t)\ dt$$

The digital PWM with sawtooth carrier has the characteristics shown in Figure 4.11.

- Reference value > timer count value – OFF
- Timer count value is zero – ON.

Figure 4.11: The digital PWM with sawtooth carrier.

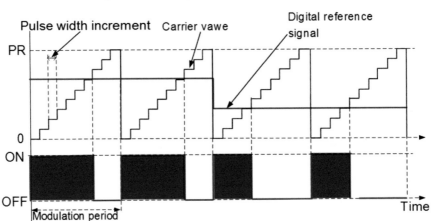

The digital PWM with triangular carrier has the characteristics depicted in Figure 4.12.

- Reference value > timer count value – ON
- Reference value < timer count value – OFF.

Figure 4.12: The digital PWM with triangular carrier.

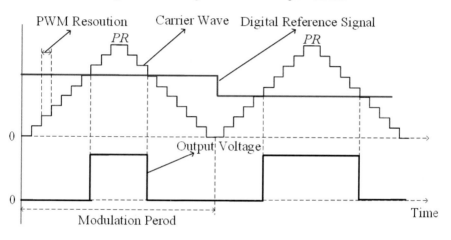

How can the PWM switching frequency be calculated?

$$PWM_{period} = PR.T_{CY}.TMR_{prescale} \tag{4.2}$$

$$F_{PWM} = PWM_{frequency} = \frac{1}{PWM_{period}}. \tag{4.3}$$

- PR – timer period register value
- T_{CY} – instruction clock cycle (T_{CY} = 1/CPU clock frequency)
- TMR_prescale – timer clock prescale value (1, 8....256).

The PWM resolution has the following characteristics:

- PWM resolution indicates the number of different widths which the output pulse can have and is defined with the following equation:

$$PWM_{resolution} = log_2 \left(\frac{CPU \ Clock \ _{Frequency}}{F_{PWM}.TMR_{prescale}} \right) \quad (bits). \tag{4.4}$$

- For example, a 10-bit resolution PWM can have 1024 different widths for the pulses, the higher the PWM resolution, the more precise the approximation of a given reference, as illustrated in Figure 4.13.

Figure 4.13: The more different widths for pulses, the higher the PWM resolution.

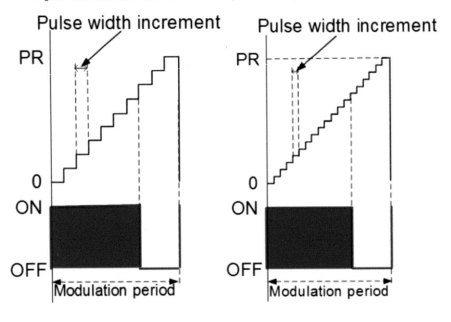

The duty cycle properties are as follows:

- The duty cycle is the compare value divided by the period:

$$DutyCycle = \frac{PWM\_CV}{PWM\_period}.$$ (4.5)

- The duty cycle can vary between 0 and 1 (in our case when it is 0 the I/O pin stays permanently in 0 V while when it is 1 it is permanently 3.3 V).

For generating a sinusoidal voltage by using PWM we the following:

- The timer used for the PWM can have only positive values; the reference analogue signal should also be shifted and scaled into the positive range, for example take a look at Figure 4.14.

Figure 4.14: Sinusoidal voltage generation by using PWM.

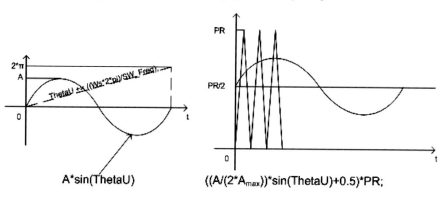

A*sin(ThetaU)   ((A/(2*A~max~))*sin(ThetaU)+0.5)*PR;

How can the negative output voltage be generated?

(1)   When the sine is in the positive range the upper switch is on for a longer time, when the negative part is generated, the bottom switch is going to be on for a longer time, as shown in Figure 4.15.

Figure 4.15: Generating negative output voltage.

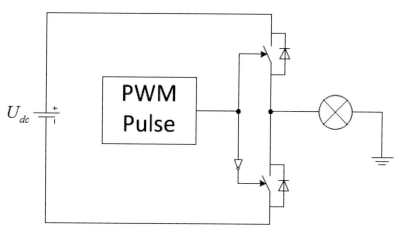

Figure 4.16: The PWM filtering.

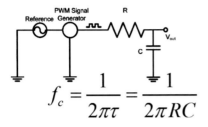

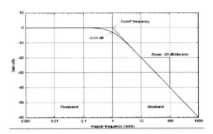

$$f_c = \frac{1}{2\pi\tau} = \frac{1}{2\pi RC}$$

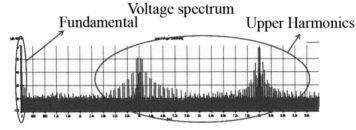

For eliminating the upper harmonics of a PWM, we use a low pass RC filter for PWM filtering, as depicted in Figure 4.16.

The PWM module from 320F28335 has the following characteristics, as illustrated in Figure 4.17:

- 6 PWM modules
- 16 bit time-based counter
- Two independent outputs for each (A and B)
- Different action qualifier settings
- Dead band generation
- ADC start event generation.

Also, the internal structure of the PWM unit is depicted in Figure 4.18.

The register shadowing has the following characteristics:

- To avoid randomly changing of the content of a register which are containing hardware related information
- Example: Randomly changing the period register can cause unexpected long delays:

Figure 4.17: The PWM module from 320F28335.

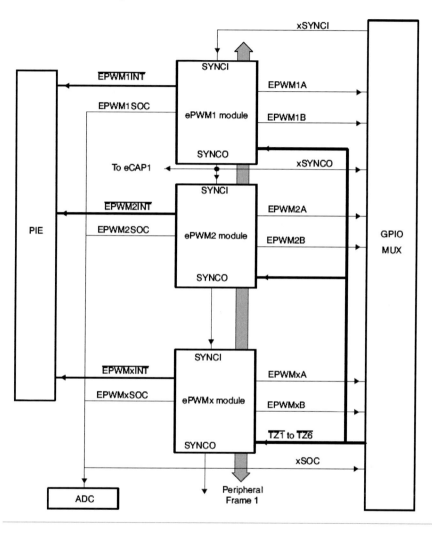

- o At the time when the counter value is 100 and we load to the PR register a value smaller 100, the counter will match the PR only after overflow, which will take a long time.
- o With a shadow mechanism the new value of the PR is loaded only when the counter is 0 or PR.

Figure 4.18: The internal structure of the PWM unit.

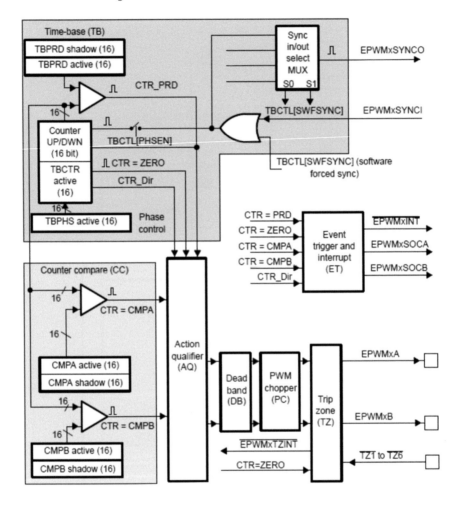

The dead band generator has the following properties:

- Due to parasitic inductances and capacitance, after a switch of a half bridge is closed/opened, a DC short circuit can appear, which can be avoided by waiting a short time, as illustrated in Figure 4.19.

Figure 4.19: The dead band generator.

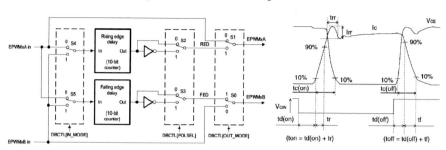

The interrupt generation has the following characteristics:

- Each module can generate an interrupt on the top and/or in the bottom of the carrier wave

  - In motor control a current sampling on this even is started because it is happening when the current is close to the average value.

- Interrupt can be generated at instance when the counter value is equal with the compare register value

  - The current sampling at this instance can cause noisy measurement.

For setting up the PWM module do the following steps:

- Setting of the GPIO register:

  - Enable the pull up resistors for the I/O pin
  - Select the PWM unit to control the I/O pin.

- Setting of the PWM module

  - Setting of the period register (switching frequency)
  - Shadow registers
  - Timer counting mode setting
  - Setup the action when the compare value is reached.

- Interrupt settings

- o Enable/disable
- o Set the event when interrupt is generated.

As a summary of digital to analog converters, we examined the following items:

- Parameters of DAC
- Types of DAC
- PWM pulse generation methods
- PWM module from 320F28335.

4.3 Lab4

We will do the following tasks in this section:

- Task 1. Rewrite the Set_SW_Freq(float sw_freq) function to calculate the value of the period register in a way to have the same switching frequency like the value in the input parameter float sw_freq.
- Task 2. Calculate the value of "ScaleDA" to set the average voltage proportional to the position of the slider, if the position of the slider is 300, to have 3 V, if 150, to have 1.5 V.
- Task 3. Calculate the compare register value (CVU_BUFF) in such a way to have the peak-to-peak amplitude for the sine wave in volts, receiving the amplitude value through the "Ampl_SL" slider.

In this lab, you will perform a digital to analog (DAC) conversion using the PWM technique. To implement the digital to analog conversion, do the following steps:

- (1) Exclude Lab3.c from the project.
- (2) Copy and add project pwm.c, pwm.h, slider.gel and Lab4.c files to the Lab4.c.
- (3) Connect the probes of the oscilloscope to CH_0 (1st) and to the filtered IOO is the (4th).
- (4) Ensure #define Separate_Volt_Level is uncommented and #define Sine_Volt is commented in Lab4.c, as you can see in the code:

/*

* Lab4.c

*/

#include "IO.h"

#include "DSP2833x_Device.h" // DSP2833x Headerfile Include File

```c
#include "DSP2833x_Examples.h"

#include "pwm.h"

#include "constants.h"

#define Separate_Volt_Level

//#define Sine_Volt

interrupt void ePWM1_isr(void);

// Global Variables

//slider variables

int Speed=1, change = 0, SW_freq_Slider=5000;

int DutyU;//duty from slider

//PWM variables

unsigned short CVU_BUFF; // compare value register "U, V, W"

unsigned short PR; // period register in rising of PWM triangular

float SW_Freq = 5000.0; // Switching frequency

float ScaleDA = 20.0;

//sine generation

float ThetaU = 0.0, Ws = 40.0;

int Amplitude;

void main(void){

InitSysCtrl();

DINT;//disable interrupts

InitPieCtrl();

IER = 0x0000; //disable all interrupts

IFR = 0x0000; //clear all interrupts flags

InitPieVectTable();

    // Interrupts that are used in this //example are re-mapped to
```

```
EALLOW; // This is needed to write to EALLOW protected register

*(Uint32 *) &PieVectTable.EPWM1_INT = (Uint32)ePWM1_isr;

EDIS; // This is needed to disable write to EALLOW protected registers

DINT;

EALLOW;

SysCtrlRegs.PCLKCR0.bit.TBCLKSYNC = 0;//synchronize the three PWM
clocks

EDIS;

PWM_HW_Init();

IER = M_INT1|M_INT3;

// Enable ePWM1 in the //PIE: Group 3 //interrupt 1

PieCtrlRegs.PIEIER3.bit.INTx1 = 1;

PieCtrlRegs.PIEIER3.bit.INTx2 = 0;

PieCtrlRegs.PIEIER3.bit.INTx3 = 0;

EINT; // Enable Global interrupt INTM

ERTM; // Enable Global realtime interrupt DBGM

Start_PWM1_INT();

while(1){

if(change != 0){

if((float)SW_freq_Slider!=SW_Freq){

SW_Freq = (float)SW_freq_Slider;

PR = Set_SW_Freq(SW_Freq);

}

#ifdef Separate_Volt_Level

CVU_BUFF = (unsigned short)((float)DutyU*ScaleDA);

#endif
```

```
#ifdef Sine_Volt

Ws = (float)Speed;

#endif

change = 0;

}

}

}//void main(void){

interrupt void ePWM1_isr(void)

{//interrupt generated when counter is 0

#ifdef Sine_Volt

ThetaU = ThetaU + ((Ws*TWO_PI)/SW_Freq);

ThetaU = fmod(ThetaU, TWO_PI);//limit //theta between 0-2pi

CVU_BUFF=(((float)Amplitude*50.0)*sin(ThetaU)+2500.0);//why 2500???, since
2500 = 0.5*5000

#endif

PWM_Update();//top update

//IO2_Toggle();

// To receive more interrupts ePWM modul

EPwm1Regs.ETCLR.bit.INT = 1;

// To receive more interrupts from this PIE //group, acknowledge this interrupt

PieCtrlRegs.PIEACK.all = PIEACK_GROUP3;

}
```

(5) Build/run the code.
(6) Load the slider.gel file like you did in the previous lab.
(7) By moving the slider the duty ratio will be changed, on the filtered an approximately DC voltage can be obtained between 0 and 3.3 V, as shown in Figure 4.20.

Figure 4.20: Variable duty ratio rectangular and its filtered waveforms.

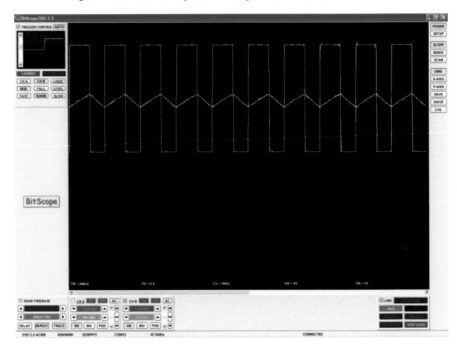

(8)   Task 1. Rewrite the Set_SW_Freq(float sw_freq) function in order to return a value for the period register which will result in the same switching frequency like the input parameter float sw_freq. One solution could be:

void Set_SW_Freq(float sw_freq) {

SW_Freq = sw_freq;

}

Notice:

- That by increasing the switching frequency the ripple is reduced.
- The switching frequency cannot be reduced below 1100 Hz. Do you have any explanation?

(9)   Task 2. Calculate the value of "ScaleDA" to set the average voltage proportional to the position of the slider, for example if the position of the slider is 300, to have 3 V at the output of the filter, if 150, to have 1.5 V.

ScaleDA = 20*3/300 = 20*1.5/150 = 0.2

(10)   Comment #define Separate_Volt_Level and uncomment and #define Sine_Volt.

(11)   Build/run the code.

(12)   On the scope a sinusoidal signal will appear as in Figure 4.21.

Figure 4.21: A sinusoidal signal on the scope.

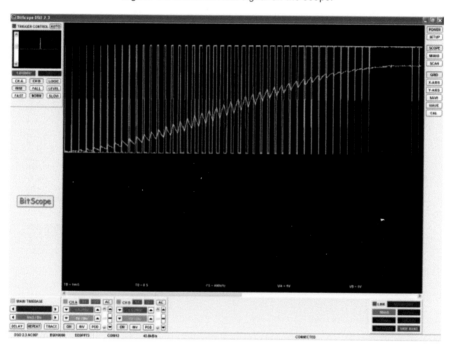

(13)   By changing the value of Ws, the frequency of the sinusoidal signal will be changed, by changing the amplitude variable, the amplitude of the sinusoidal signal will be changed.

(14)   Task 3. Calculate the compare register value (CVU_BUFF) in such a way to have the peak-to-peak amplitude for the sinus in volts received from the "Ampl_SL" slider. The solution would be to define a gel file named Ampl_SL.gel as:

```
menuitem "Ampl_SL"

slider Ampl_SL(0,50,1,1,Ampl_SL){

Amplitude = Ampl_SL;

}
```

(15)   Optional: Set up a second PWM module and generate a sinusoidal signal with 90-degree phase shift compare to the first sinusoidal signal. The solution will be to change the variable ThetaU as:

ThetaU = ThetaU + ((Ws*TWO_PI)/SW_Freq) + PI/2;

This ends Lab4.

5

Analog to Digital Conversion

5.1 Introduction

In the previous chapter, we examined the following items:

- Parameters of DAC.
- Types of DAC.
- PWM pulse generation.
- PWM module from 320F28335.

The agenda of analog to digital conversion (ADC) is as follows:

- What is ADC?
- Requirements of ADC.
- ADC module of TMS320F28335.

The sensing quantity has the following characteristics:

- Generally, all the quantities (voltage, current, temperature, etc.) can be converted into scaled voltage signal by using sensors.
- Example: Current signals can be measured with the help of a measuring resistor, knowing the resistor value and by measuring the voltage drop on the resistor. Ohm's law gives the current through the resistor.
- Temperature sensors can also be resistors, which changes their resistance as a function of the temperature; by applying a constant

current through the resistor and measuring the voltage drop on it the resistance can be calculated which is proportional to the temperature.

5.2 The Realization of Analog to Digital Conversion (ADC)

What is ADC?

- ADC converts an analogue, continuous signal at its input into a digital discrete number at its output.
- The ADC is a mixed-signal device.

The example of an ADC is as follows:

- For a 3-bit ADC, there are eight possible output codes.
- In this example, if the input voltage is 5.5 V and the reference is 8 V, then the output will be 101 (6 V), as shown in Figure 5.1.

Figure 5.1: An example of an ADC.

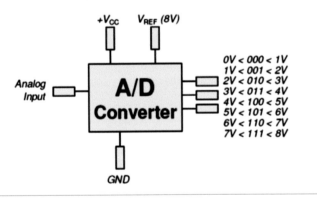

The transfer function of an ADC at different resolution is depicted in Figure 5.2.

The ADC resolution has the following characteristics. The resolution is the number of discrete values it can produce over the range of analogue values:

- It can be given in bits: for example, a 12-bit resolution ADC can produce $2^{12} - 1 = 4095$ discrete digital values of its full-scale analogue input voltage.

Figure 5.2: The transfer function of ADC at different resolution.

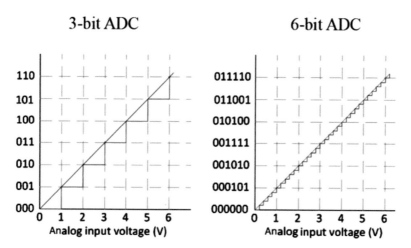

- It can be given in volts: For example, with a full-scale input of 0–3 V, it will have a resolution of 3/4095 = 0.00073 V.

The periodic sampling includes the properties illustrated in Figure 5.3.

Figure 5.3: The periodic sampling properties.

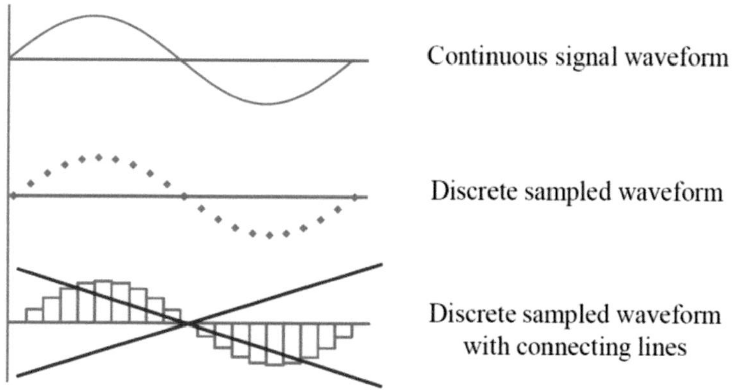

Continuous signal waveform

Discrete sampled waveform

Discrete sampled waveform with connecting lines

The theory of aliasing and anti-aliasing filters is shown in Figure 5.4.

Figure 5.4: The theory of aliasing and antialiasing filters.

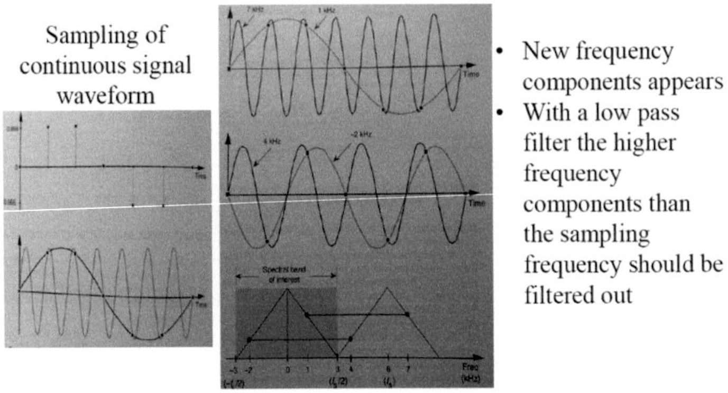

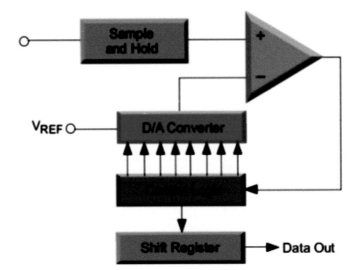

Sampling of continuous signal waveform

- New frequency components appears
- With a low pass filter the higher frequency components than the sampling frequency should be filtered out

The typical scheme for conversion is depicted in Figure 5.5.

Figure 5.5: The typical scheme for conversion.

The sample and hold circuit characteristics are as follows:

- The purpose of the S/H circuit is to keep the voltage level constant during the conversion (mainly a capacitor).

- The settling time (time needed to charge up the S/H capacitor) limits the maximum input frequency.

The typical ADC input circuit is illustrated in Figure 5.6.

Figure 5.6: The typical ADC input circuit.

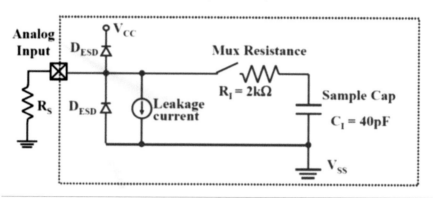

The typical ADC input circuit has the following properties:

- Input internal impedance is relatively low.
- A high impedance source increases sample capacitor charging time.
- Rise time of voltage on $T_I \sim (R_S + R_I) * C_I$.

The successive approximation for an ADC is shown in Figure 5.7 and has the following characteristics:

Figure 5.7: The successive approximation for an ADC.

(1) MSB of DAC input is set to "1" (half of DAC output range).
(2) Test if DAC output is higher than analog input. If higher, MSB = 0, else MSB = 1.
(3) Repeat 1 and 2 with MSB-1.

The summary of the characteristics of the ADC are as depicted in Figure 5.8.

The full conversion time is a two-step process:

(1) Charging the sampling capacitor.
(2) Disconnect C_{hold} from the input pin and start the A/D conversion.

Figure 5.8: The summary of the characteristics of the ADC.

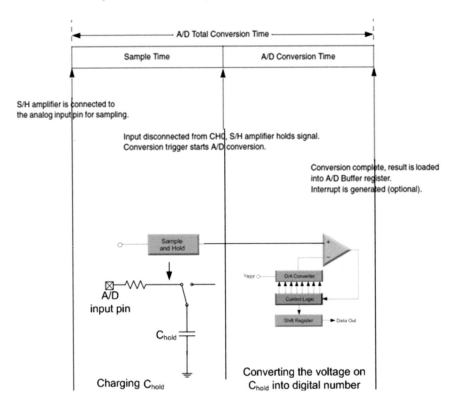

The 320F28335 ADC module has the following characteristics:

• 12-bit resolution ADC core.
• Sixteen analog inputs (range of 0 to 3 V).

- Two analog input multiplexers.
- Up to eight analog input channels each.
- Two sample/hold units (for each input mux.).
- Sequential and simultaneous sampling modes.
- Auto sequencing capability – up to 16 auto conversions.
- Two independent 8-state sequencers.
- "Dual-sequencer mode".
- "Cascaded mode".
- Sixteen individually addressable result registers.
- Multiple trigger sources for start-of-conversion.
- S/W – software immediate start.
- ePWM 1–6.
- GPIO XINT2.
- ePWM triggers can operate independently in dual-sequencer mode.

For example, the sequencer "start/stop" operation is illustrated in Figure 5.9.

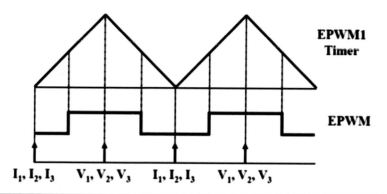

Figure 5.9: The sequencer "start/stop" operation.

The system requirements are as follows:

- Three auto conversions (I_1, I_2, I_3) off trigger 1 (timer underflow).
- Three auto conversions (V_1, V_2, V_3) off trigger 2 (timer period).

EPWM1 and SEQ1 are used for this example with sequential sampling mode.

The ADC module block diagram for the cascaded mode is shown in Figure 5.10.

Figure 5.10: The ADC module block diagram for cascaded mode.

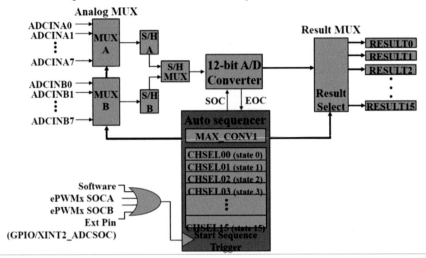

The ADC module block diagram for dual-sequencer mode is depicted in Figure 5.11.

Figure 5.11: The ADC module block diagram for dual-sequencer mode.

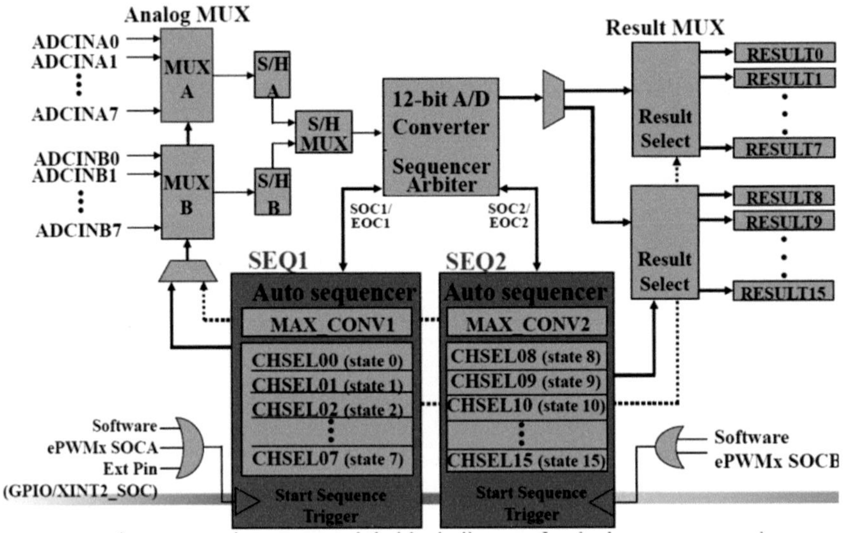

The ADC conversion result buffer register is illustrated in Figure 5.12.

Figure 5.12: The ADC conversion result buffer register.

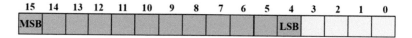

With analog input 0V to 3V, we have:

analog volts	converted value	RESULTx
3.0	FFFh	1111|1111|1111|0000
1.5	7FFh	0111|1111|1111|0000
0.00073	1h	0000|0000|0001|0000
0	0h	0000|0000|0000|0000

How do we read the result of ADC conversion? The procedure for reading the result of an ADC conversion is shown in Figure 5.13.

Figure 5.13: Reading the RESULT0 register example.

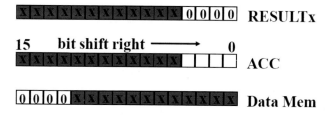

The program codes for reading the RESULT0 register of ADC conversion are as below:

```
#include "DSP28xxx_Device.h"

void main(void)

{

Uint16 value;// unsigned

value = AdcRegs.ADCRESULT0 >> 4;

}
```

The ADC clocking block diagram example is depicted in Figure 5.14.

Figure 5.14: The ADC clocking block diagram example.

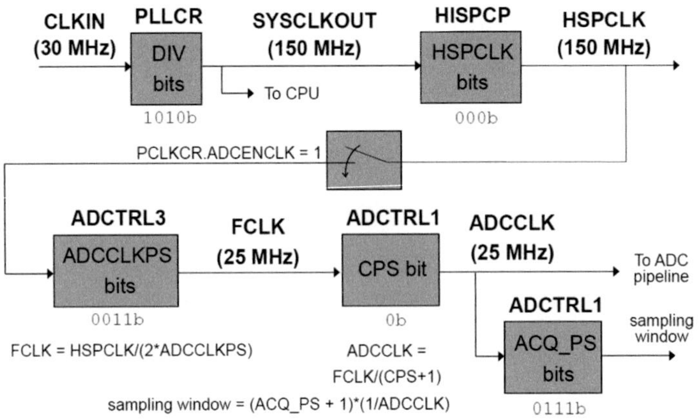

The important note is that the ADCCLK can be a maximum of 25 MHz; a summary of the ADC conversion that we examine in this chapter is as follow:

- The analog signal should be scaled down to the input voltage range of the ADC module.
- The reference voltage and the number of the bits in the ADC will give the resolution of the ADC.
- The start of conversion (SOC) signal can be generated either by software or by the internal/external hardware of the DSC.
- The gain factor of the sensing quantity should be calculated in the control program.
- The ADC can be configured in cascade or dual-sequencer mode.

5.3 Lab5

In this section, we will carry out the following tasks:

- Acquire a sinusoidal signal through the ADC port.
- Toggle a GPIO pin after each AD conversion.
- Change the speed of the AD conversion (change the ADC clock).
- Determine when the signal is below a threshold (i.e., near zero); if the signal is below that threshold, clear one GPIO pin to be 0 else (measured signal is above the threshold) set the GPIO to 1.
- Scale the measured signal to get the same amplitude values as you have on the scope (signal generator) screen.
- Take out the dc offset and at zero crossing toggle a GPIO pin.

94

In this lab, you will perform an analog-to-digital conversion (ADC). To implement the analog to digital conversion, do the following steps:

(1)   Generation of a signal using Picoscope:

      (a)   Connect the Picoscope to the PC and start its software.

      (b)   Hook the CHA and AWG scope cables together.

      (c)   The generated signal should be seen on the scope screen of the PC.

(2)   Generate a 1 V peak-to-peak sinusoidal waveform with a dc bias of 0 V and a frequency of 10 Hz. Your sine wave should oscillate between −0.5 and 0.5 V. Note that between the BNC connector and the ADC channel there is a voltage shifter. The ADC input channels are designed for signals ranging between 0 and 3 V. Values less than that will be read as zero, while values larger than that will produce a full-scale conversion result. Of course, if you put in values that are too far out of the range, you may damage the DSP board. Do not change the amplitude parameters of the generated signal when the demo board is connected to avoid the damage of the A/D channel. You should see the image shown in Figure 5.15.

Figure 5.15: The sine wave oscillating between -0.5 and 0.5 volts.

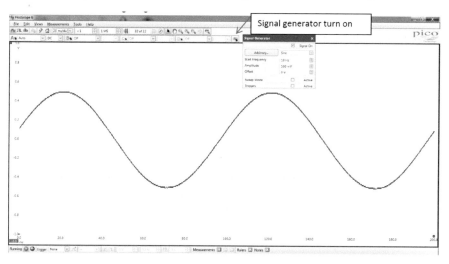

(3)   Connect the AWG channel to ADCB0 input on the experimental setup.

(4)   Exclude Lab4.c file from the project.

(5)   Copy the files adc.c, adc.h, and Lab5.c to the project.

The adc.c codes:

```
//**********************************************************
```

// Lab5 - Analog to Digital Conversion

```
//**********************************************************
```

```
#include "adc.h"
void Init_Adc(){
    extern void DSP28x_usDelay(Uint32 Count);
    EALLOW;
SysCtrlRegs.PCLKCR0.bit.ADCENCLK = 1;
ADC_cal();
EDIS;
    AdcRegs.ADCTRL3.all = 0x00E0; // Power up bandgap/reference/ADC
    DELAY_US(ADC_usDELAY); // Delay before converting ADC
// Specific ADC setup for this example:
AdcRegs.ADCMAXCONV.all = 0x00; // //Setup 1 conv's on SEQ1
AdcRegs.ADCCHSELSEQ1.bit.CONV00 = 0x8; // First A/D channal
    AdcRegs.ADCTRL2.bit.EPWM_SOCA_SEQ1 = 0; // Disable SOCA from
ePWM to //start SEQ1
    AdcRegs.ADCTRL2.bit.INT_ENA_SEQ1 = 0; // Disable SEQ1 interrupt
    AdcRegs.ADCTRL2.bit.INT_MOD_SEQ1 = 0; // Have interrupt only once //per
sequence
    AdcRegs.ADCTRL3.bit.ADCPWDN = 1; //power the AD circuits
    AdcRegs.ADCTRL1.bit.CONT_RUN = 1; // Setup continuous run
    AdcRegs.ADCTRL1.bit.ACQ_PS = 0x4; //clock prescaler
    AdcRegs.ADCTRL3.bit.ADCCLKPS = 0x1; //ADCCLK
    AdcRegs.ADCTRL1.bit.CPS = 0; //no extra division
  AdcRegs.ADCTRL2.all = 0x2000;
}
```

The adc.h codes:

```
//*********************************************************
// Lab5 - Analog to Digital Conversion
//*********************************************************
#ifndef ADC_H
#define ADC_H
#include "DSP2833x_Device.h"
#include "DSP2833x_GlobalPrototypes.h"
#include "DSP2833x_EPwm_defines.h"
#include "DSP2833x_Examples.h"
extern void DSP28x_usDelay(Uint32 Count);
extern void Init_Adc();
#define CLOCK_FREQUENCY ((double)150e6) // 150 MHz clock
#define      CLOCK_PERIOD      ((double)6.667e-9      )      //
//1/CLOCK_FREQUENCY clock period
// ADC start parameters
#define ADC_usDELAY 5000L
#define      ADC_MODCLK      0x3      //      HSPCLK      =
SYSCLKOUT/2*ADC_MODCLK = //150/(2*3) = 25.0 MHz
#define ADC_CKPS 0x1 // ADC module clock = HSPCLK/2*ADC_CKPS =
//25.0MHz/(1*2) = 12.5MHz = 16 ADC clocks
#define BUF_SIZE 200 // Sample buffer size
#endif
```

The Lab5.c codes:

```
//*********************************************************
// Lab5 - Analog to Digital Conversion
//*********************************************************
#include "IO.h"
```

```c
#include "DSP2833x_Device.h" // DSP2833x Headerfile Include File
#include "DSP2833x_Examples.h"
#include "adc.h"
// Global Variables
float SampleTable[BUF_SIZE];
float AD_Value;
int pointer=0, i;
void main(void){
    InitSysCtrl();
EALLOW;
    SysCtrlRegs.HISPCP.all = ADC_MODCLK; // set the ADC clock
    EDIS;
    DINT;//disable interrupts
    InitPieCtrl();
    IER = 0x0000; //disable all interupts
    IFR = 0x0000; //clear all interrupts flags
    InitPieVectTable();
IO_Init();
Init_Adc();
while(1){
    while (AdcRegs.ADCST.bit.INT_SEQ1== 0) {} // Wait for interrupt
    AdcRegs.ADCST.bit.INT_SEQ1_CLR = 1;//clear the interrupt flag
    AD_Value = ((float)((AdcRegs.ADCRESULT0>>4));
if(pointer++>BUF_SIZE) pointer=0;
    SampleTable[pointer] = AD_Value;
}
}//void main(void){
```

(6)   This program acquires data from channel B0 of the ADC.

(7)   When you run the program, you can look at the Watch Window like you did in Lab1 to see how your variables are changing. In this lab, the variable of interest is actually an array of values acquired by the ADC, called "SampleTable". For this lab, we're going to use the graphing capabilities of CCS. When the program has been built, loaded, and is ready to be run, select Target/Advanced/Enable Silicon Real Time Mode (you can also press the clock pictogram). Then select Tools/Graph/Single Time. You will then be presented with a table of graph properties to be filled in, such as those depicted in Figure 5.16.

Figure 5.16: The table of graph properties.

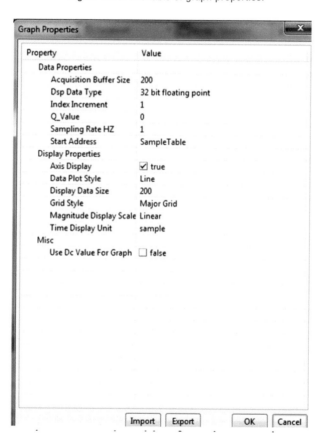

| (8)   The graph window will look that illustrated in Figure 5.17.

Figure 5.17: The graph window.

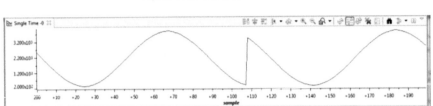

(1)   When the program runs, a random signal will be on the graph due to the fact that the refresh rate is much slower than the frequency of the measured signal.

(2)   The running program is not displaying a full period of the measured signal, because there is no triggering and the graphical refresh is much slower than the refresh of the buffer.

(3)   To see one measurement period the program should be stopped.

(4)   Your task in this lab is to perform the following:

    (a)   Scale the measured signal to get the same amplitude values as you have on the scope screen.
Solution: From Figure 5.17, we should map 2×10^2 to -0.5 and 3.3×10^3 to $+0.5$. Therefore, the relationship equation will be as in Equation (**5.1**).

$$y = -0.5 + (x - 200) \times \frac{1}{3100} \qquad (5.1)$$

AD_Value = ((float)((AdcRegs.ADCRESULT0>>4-200)*1/3100-0.5));

    (b)   Toggle a GPIO pin after each AD conversion.
Solution: We add IO.c and IO.h files to the project and also add the following code to the Lab5.c file after the above code (AD_Value equation):
I00_Toggle();

    (c)   Calculate the ADC clock which is used in the program.
Solution:
HSPCLK = SYSCLKOUT/2*ADC_MODCLK = 150/(2*3) = 25.0 MHz
FCLK = HSPCLK/2*ADC_CKPS = 25.0MHz/(1*2) = 12.5MHz
ADC module clock = FCLK/(CPS+1)
CPS = 0
16 ADC clocks = FCLK = 12.5MHz

(d)   Change the speed of the A/D conversion (change the ADC clock).
      Solution: We can change the ADC clock with changing one or multiple variables below.
      ADC_MODCLK
      ADC_CKPS
      CPS
      Note that it also can be changed with varying the following code:
      AdcRegs.ADCTRL3.bit.ADCCLKPS = 0xf; //ADCCLK

(e)   Determine if the signal at any time is below a threshold (i.e., nearing zero), if the signal is below that threshold, clear one GPIO pin to be 0 else (measured signal is above the threshold) set the GPIO to be in 1 as in Figure 5.18.

Figure 5.18: Determining the signal threshold.

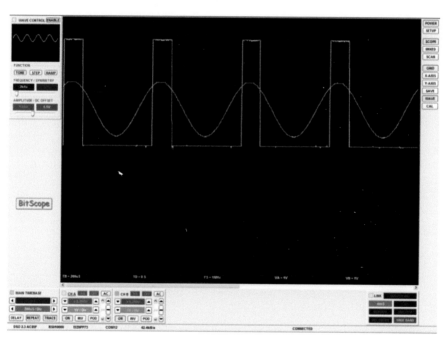

Solution: After the code was written in the above (a) section we add the following codes:

```
if (AD_Value < 0.0) {

IO0_OFF();

}
```

if (AD_Value > 3.0) {

IO0_ON();

}

(f) Take out the dc offset and toggle a GPIO pin when the value of the sinusoidal signal is above a threshold, as shown in Figure 5.18.
The solution is as below.
if (AD_Value > 3.0) {
IO0_Toggle();
}
You are now able to acquire a signal through the ADC, perform an operation, and output a control signal based on that input signal.

6

Code Generation Using Simulink, PSIM, and PLECS

6.1 Introduction

In the previous chapter, we examined the following items for analog to digital conversion (ADC):

- ADC converts an analogue, continuous signal at its input into a digital discrete number at its output.
- All the quantities (voltage, current, temperature etc.) can be converted into a voltage signal by using sensors.
- The voltage has to be scaled in order to not exceed the input voltage range of the ADC module (0–3 V in the case of 320F28335).
- The reference voltage and the number of bits in ADC will give the resolution of ADC (accuracy).
- Start of conversion (SOC) signal can be generated either by software or by internal/external hardware of the DSC.
- ADC can be configured in cascade or sequential mode.

In this chapter, we will examine the code generation using Simulink, PSIM and PLECS. Simulink, PSIM and PLECS, all offer code generation capabilities for the TMS320F28335 microcontroller. Simulink Embedded Coder is a tool that allows generating C code from Simulink models. PSIM and PLECS are simulation tools specially designed for power electronic systems. They also support code generation for the DSP TMS320F28335. After generating the code, you can import it into your TMS320F28335 development environment and build and flash it onto the microcontroller.

6.2 Code Generation Using Simulink

In this section, we are going to perform the sinusoidal pulse width modulation with the help of an analog to digital converter (ADC) and the pulse width modulation modules of the DSP processor. So, in this section we are going to convert the MATLAB Simulink model into the Code Composer Studio (CCS). So, some of the steps we are going to follow are the kind of steps we need to follow to convert the MATLAB Simulink model into the Code Composer Studio (CCS). For this we need the ADC block, zero order hold gain and three gains as shown in Figure 6.1.

Figure 6.1: The MATLAB Simulink model.

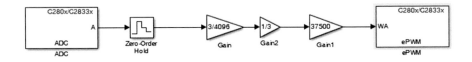

That is, the TBPRD and the ePWM block will generate the PWM pulses. So, we will see what is inside the how to select the values from the ADC block, as depicted in Figure 6.2.

So, these are the modules. There are two modules A, B, and A and B. Therefore, we can select any one and in the conversion mode we can use a sequential and simultaneous conversion for ADC. At the start of conversion with we use the software ePWM and ADCSOC. The sampling time is 5e-6 and the data type is double. For input channels we are using the one conversion. So, we are taking here the conversion number and the conversion number is ADC0, as illustrated in Figure 6.3.

This ADC will generate the digital output in the form of 4096 as it is a 12-bit ADC and zero order hold will keep the output and hold the ADC output as shown in Figure 6.4.

Now this gain is converting your ADC output into 3 V. By dividing 3 by 4096, we will obtain 3 V as the output. Additionally, we are converting the 3 V into 1 V. So, 3 V divided by 3 will give you the 1 V. So, 1 V is

Figure 6.2: The ADC block properties.

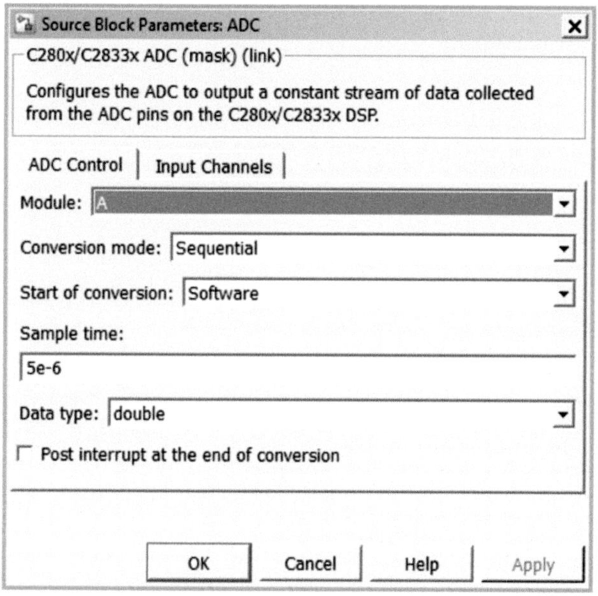

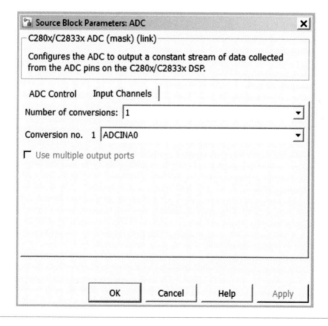

Figure 6.3: The input channels setting.

Figure 6.4: The zero-order hold tuning.

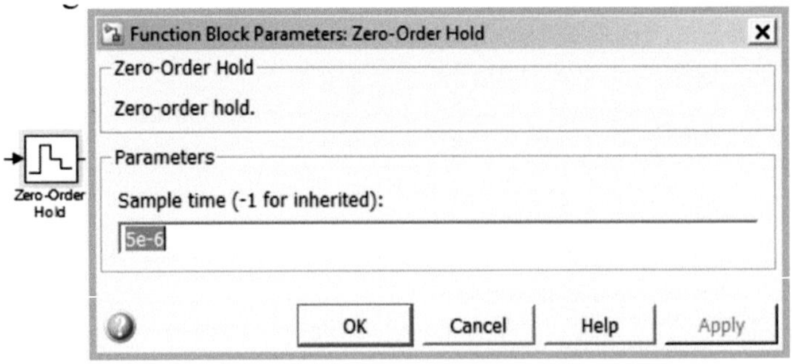

further multiplied by the TBPRD. Then, when it is 0, we will get 0, when it is 1, we will get output 37500, as depicted in Figure 6.1. We choose the ePWM1 module from the ePWM block, which offers 16 ePWM channels. As shown in Figure 6.5, we set the time period as TBPRD with a value of 37500.

Figure 6.5: The ePWM block general parameters.

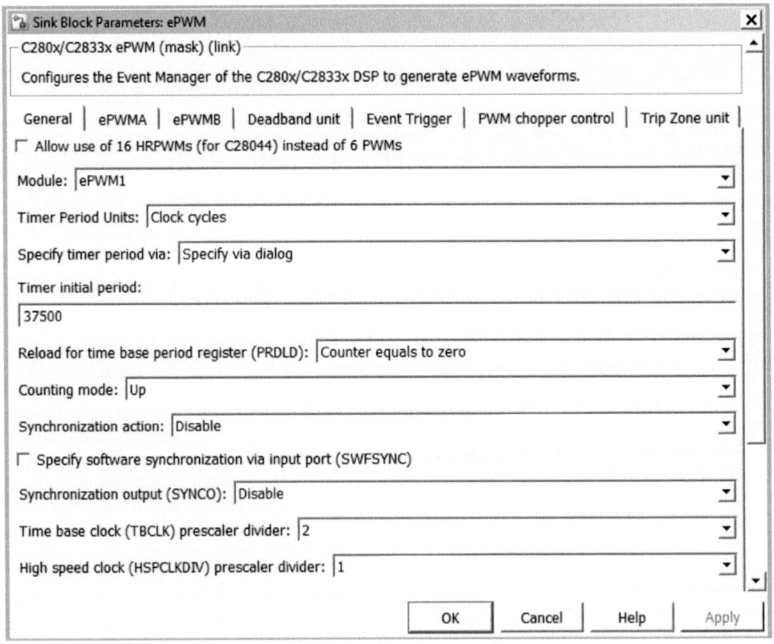

This is the 2 kHz for the 2-clock frequency:

$$period = \frac{37500}{150M/2} = 0.0005 \rightarrow frequency = \frac{1}{period} = 2\,kHz \qquad (6.1)$$

So, we are using counting mode when the counter equals zero. You can use up, down, and up down counting mode. The time base clock (TBCLK) prescaler divider is 2 and the high-speed clock prescaler divider (HSPCLKDIV) is 1. Now in this case, ePWM1A compares CMPA units using clock cycles. We can choose to specify the compares through either the input port or via dialogue. If we opt for specifying the compares via the input port, it means that we are providing the inputs to the comparators from an external source, as illustrated in Figure 6.6.

Figure 6.6: The ePWMA parameters of ePWM block.

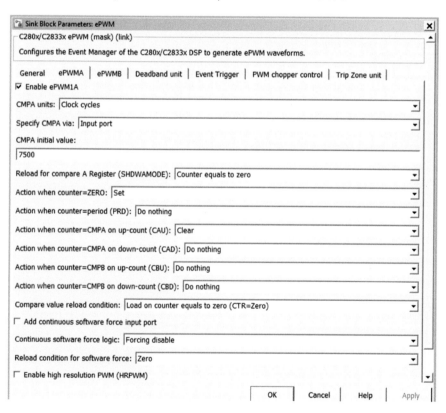

So the initial value we have kept as 7500; you can keep anything any value between 0 and the TBPRD value. In the case of SHOWAMODE, if the counter is equal to zero, we preserve the values within the SHOWAMODE register and proceed to set the output as 1.

Therefore, when the counter equals 0, we set it to 1. Similarly, when the counter reaches PTBPRD, we take no action. While the counter is counting, it is in up counter mode. In this mode, when the compare CMPA reaches a certain point, we clear that bit. We only concern ourselves with the up counting mode and do not care about down counting or any other cases. The setting order in ePWMB will be the opposite of ePWMA, as shown in Figure 6.7.

Figure 6.7: The ePWMB parameters of ePWM block.

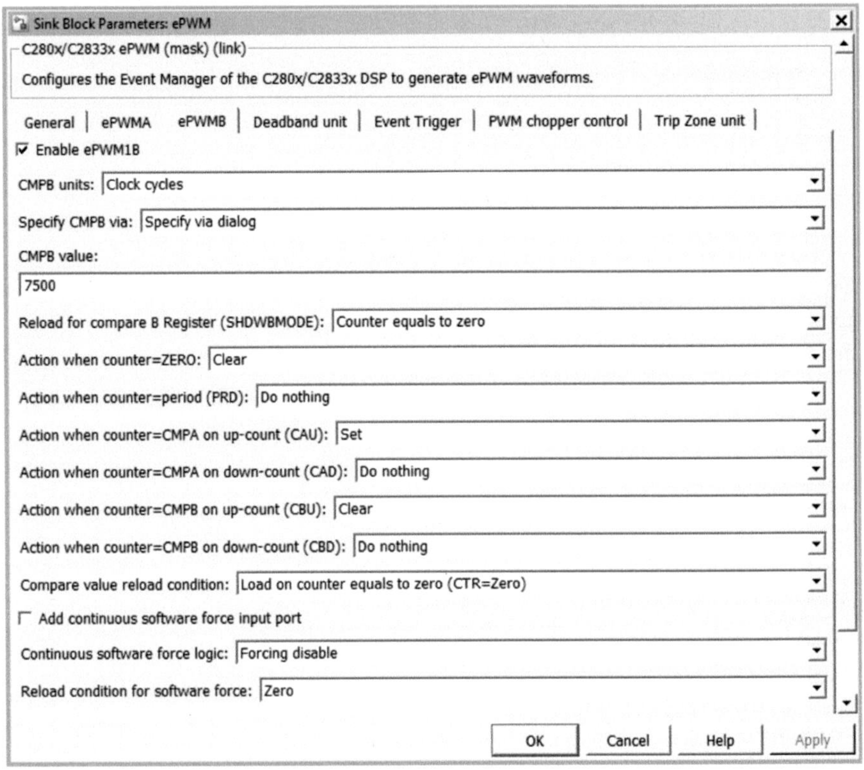

The dead-band unit is for if you want to use the inverter 1 leg so that you can use the dead band unit to generate the dead time between two legs, as illustrated in Figure 6.8.

Figure 6.8: The dead-band unit parameters.

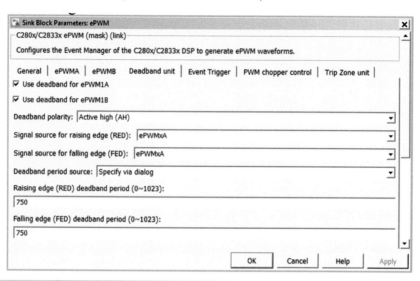

So now we have to go to the model configure parameters and hardware implementations. You have to select TI Delfino F2833X so it will automatically give you the Texas instrument and C2000 and click "okay" if you have connected the hardware to the system, as shown in Figure 6.9.

Figure 6.9: Model configure parameters and hardware implementations.

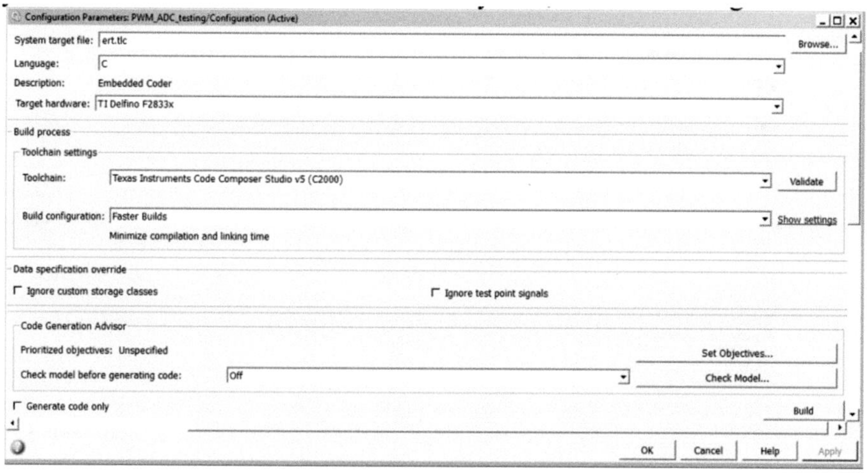

We are using TMS320F28335. Now just save it; running it will show if no error is there and pause and stop the simulations. Now deploy to the hardware as you can see in Figure 6.10.

Figure 6.10: Deploying to hardware.

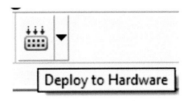

Deploying to the hardware means you are uploading and you are generating the code for this program; it will generate the code and it will generate the several libraries of several C files. So, now this is done, code has been generated successfully. Now, we will chose view diagnostic and we will see that it is successfully completed; just click on the open project in the Code Composer Studio. This will open your Code Composer Studio. Now when you click on PWM_ADC_testing it will be Active-Debug. Now click on it again and we will go to the debug section; we just run it as shown in Figure 6.11 and Figure 6.12 respectively.

Figure 6.11: Building the project.

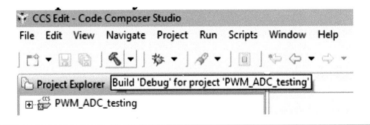

Figure 6.12: Debugging the project.

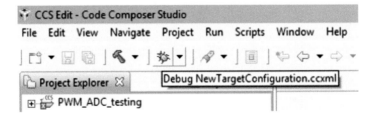

When we are running now, we see it is working fine. There is no error in the program, so now the build is finished it will be here. Now we will look at the hardware. So, this is the hardware setup we have given the 50 Hz, switching the 50 Hz sine wave from 0 to 3 V, as you can see on the screen. Here one is the PWM pulse and the other is the sine wave of the 50 Hz. The PWM frequency is 2 kHz, as depicted in Figure 6.13.

Figure 6.13: The PWM pulse and sine wave.

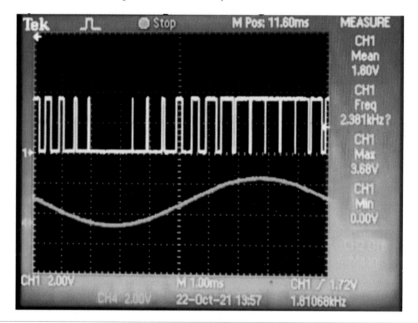

These are the PWM pulses and we are comparing the triangular wave with the sine wave to generate the PWM pulses.

6.3 Code Generation Using PSIM

PSIM will generate the necessary C code files, including the initialization code, control algorithms, and interrupt service routines. PSIM primarily focuses on simulation, although it supports code generation for TI C2000 such as the DSP TMS320F28335 microcontroller. In this section, we will generate the code using PSIM for open loop controlling of a buck converter. The overall view of the setup implementation is shown in Figure 6.14. The simulated circuit with F28335 hardware blocks in PSIM is depicted in Figure 6.15.

Figure 6.14: The overall view of the setup implementation.

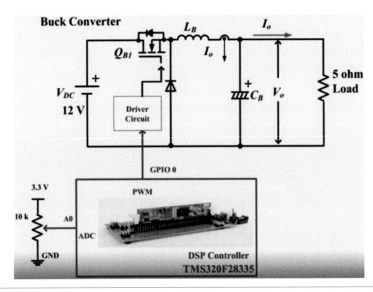

Figure 6.15: The simulated circuit with F28335 hardware blocks.

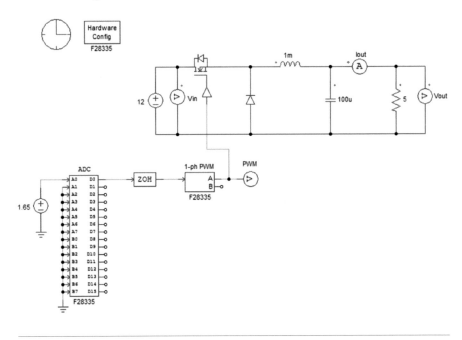

The zero-order hold parameter window is illustrated in Figure 6.16. Also, we set the peak-to-peak value to 3.3, dead time to 0.1 μs, sampling frequency to 20 kHz and one PWM (PWM A) as output in the 1-phase PWM parameter window as shown in Figure 6.17.

The A/D converter parameter settings are depicted in Figure 6.18.pc

Figure 6.16: The zero-order hold parameter window.

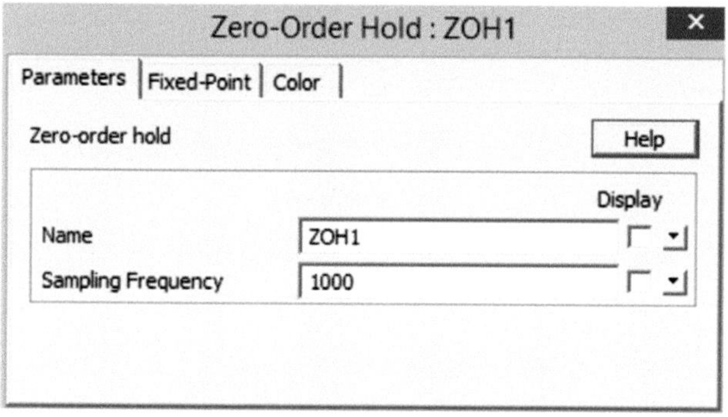

Figure 6.17: The 1-phase PWM parameter settings.

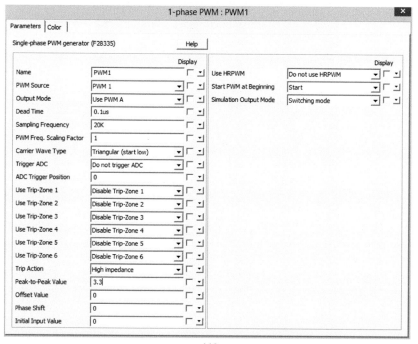

Figure 6.18: The A/D converter parameter settings.

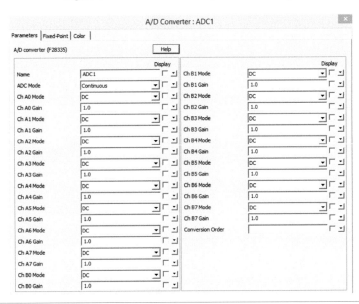

Also, we set the simulation control parameters as you can see in Figure 6.19.

Figure 6.19: The simulation control parameter settings.

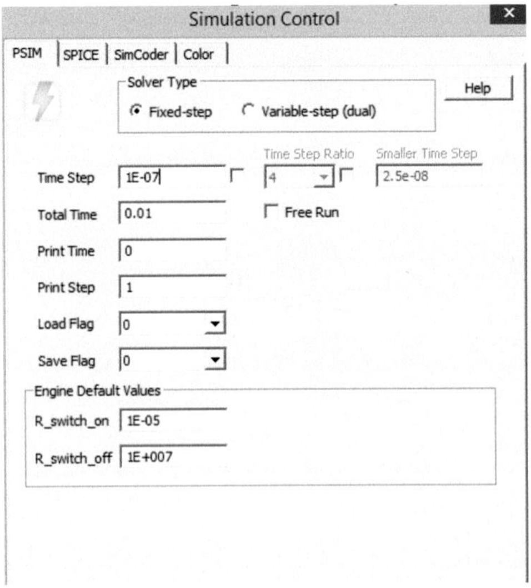

In the simulation control window and in the SimCoder tab we select F2833x as the hardware target, as illustrated in Figure 6.20. Moreover, in the hardware configuration window we select GPIO0 and GPIO1 as PWM, as shown in Figure 6.21.

Figure 6.20: The SimCoder parameter settings.

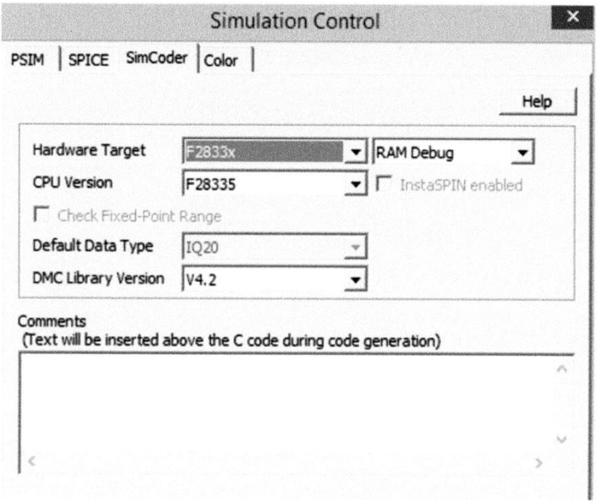

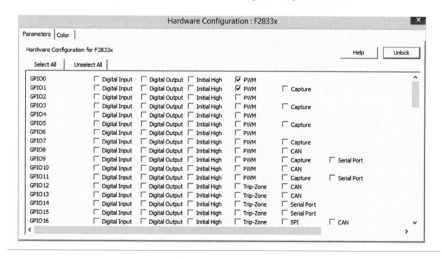

Figure 6.21: The hardware configuration parameters.

We run the simulation and the waveforms of PWM, Vin and Vout are depicted in Figure 6.22. To generate the code for the TMS320F28335 microcontroller, from the simulate tab we choose the Generate Code option, as illustrated in Figure 6.23.

Figure 6.22: The simulation waveforms of PWM, Vin and Vout.

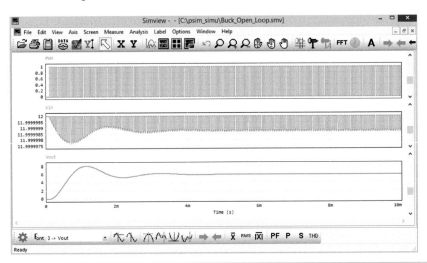

Figure 6.23: Choosing the Generate Code option.

The generated code files are shown in Figure 6.24. Then we run the Code Composer Studio (CCS) and select the generated codes folder as workspace, as depicted in Figure 6.25.

Figure 6.24: The generated code files.

Local Disk (C:) ▸ psim_simu ▸ Buck_Open_Loop (C code)

Name	Date modified	Type	Size
Buck_Open_Loop	12/12/2023 10:34 ...	C File	3 KB
Buck_Open_Loop.pjt	12/12/2023 10:34 ...	PJT File	5 KB
F2833x_Headers_nonBIOS	12/12/2023 10:34 ...	Windows Comma...	9 KB
F28335_FLASH_Lnk	12/12/2023 10:34 ...	Windows Comma...	7 KB
F28335_FLASH_RAM_Lnk	12/12/2023 10:34 ...	Windows Comma...	6 KB
F28335_RAM_Lnk	12/12/2023 10:34 ...	Windows Comma...	4 KB
passwords.asm	12/12/2023 10:34 ...	ASM File	4 KB
PS_bios.h	12/12/2023 10:34 ...	H File	21 KB
PsBiosRamF33xFloat.lib	1/24/2023 6:56 PM	LIB File	643 KB
PsBiosRomF33xFloat.lib	1/24/2023 6:56 PM	LIB File	647 KB
rts2800_fpu32_fast_supplement.lib	1/24/2023 6:56 PM	LIB File	17 KB

Figure 6.25: Selecting the generated codes folder as workspace.

From the Project menu, we select Import Legacy CCSv3.3 Project… then choose the Buck_Open_Loop.pjt file in the generated codes folder as illustrated in Figure 6.26 and Figure 6.27 respectively.

Figure 6.26: Selecting Import Legacy CCSv3.3 Project….

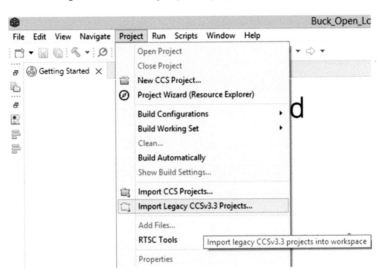

Figure 6.27: Choosing the Buck_Open_Loop.pjt file.

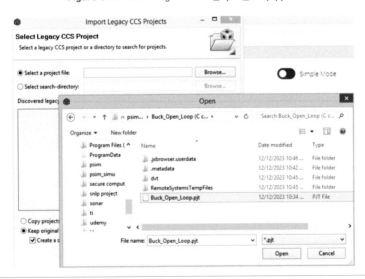

We select Keep original location for each project and then click on the Next button as shown in Figure 6.28. In the next window we select a compiler version and click on the finish button, a new window will pop up to use a common root and then click on the finish button as you can see in Figure 6.29 and Figure 6.30 respectively.

Figure 6.28: Clicking on the Next button.

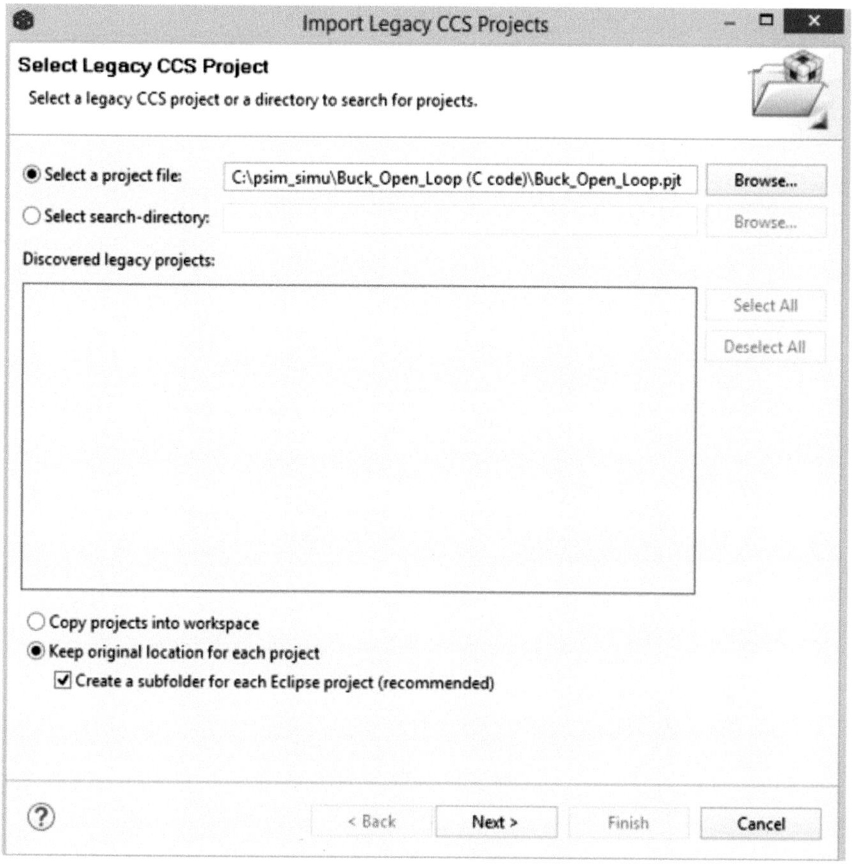

Figure 6.29: Selecting a compiler version and clicking on the finish button.

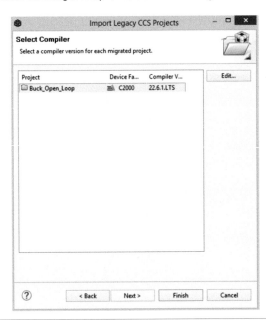

Figure 6.30: Using a common root.

We right click on the Buck_Open_Loop project and select Build Configuration section select 3 3_FlashRelease as Set Active, as shown in Figure 6.31.

Figure 6.31: Selecting the 3 3_FlashRelease as Set Active.

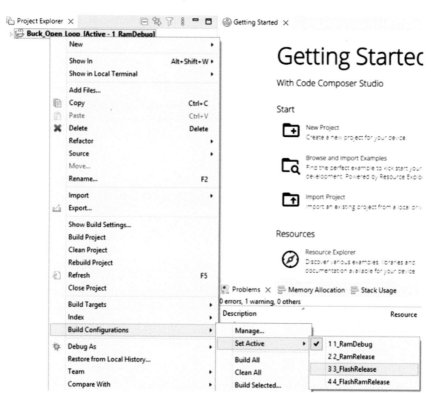

From the Run menu we select Debug and then a New Target configuration window pops up and we click on the finish button to open the general configuration about the target (F28335) window, as depicted in Figure 6.32, Figure 6.33 and Figure 6.34 respectively.

We click on the Build All item from the Project menu to build the project and we again select Debug from Run menu as illustrated in Figure 6.35 and Figure 6.32 respectively.

In the debug window we click on the Resume (F8) button to execute the project on the hardware, as shown in Figure 6.36. The generated PWM signal controlled by a potentiometer is displayed on the oscilloscope, as you can see in Figure 6.37.

Figure 6.32: Selecting Debug from Run menu.

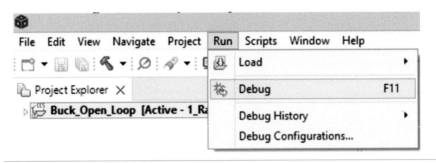

Figure 6.33: New target configuration window.

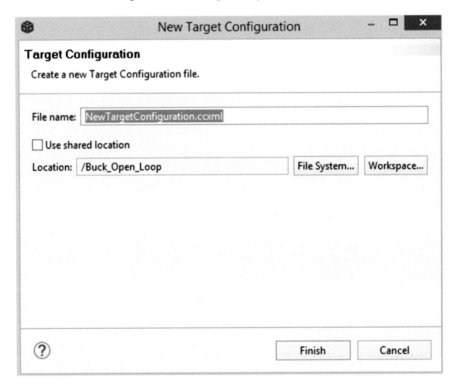

Figure 6.34: Setting the new target configuration.

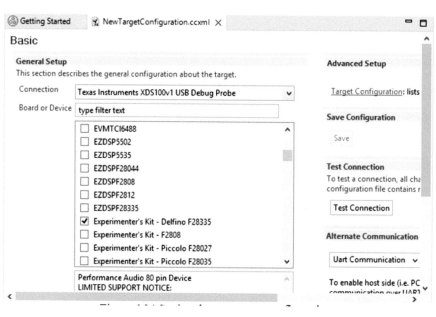

Figure 6.35: Click on the Build All item.

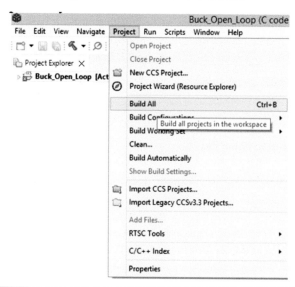

Figure 6.36: Clicking on the Resume (F8) button.

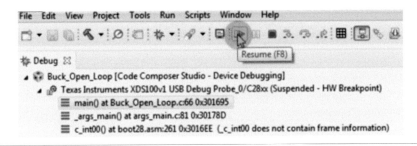

Figure 6.37: Displaying the generated PWM signal on the oscilloscope.

6.4 Code Generation Using PLECS

PLECS will generate the necessary C code and project files specific to the F28335 microcontroller. You can then import this code into a suitable integrated development environment (IDE) for the F28335 device, such as Code Composer Studio (CCS) or TI's C2000 Control Suite. Once the code is imported into your IDE, you can build and load it onto the F28335 microcontroller for execution. In this section, we will generate codes for a closed loop controlled buck converter using PLECS. The simulated circuit in the PLECS software environment is shown in Figure 6.38.

Figure 6.38: The simulated circuit in PLECS.

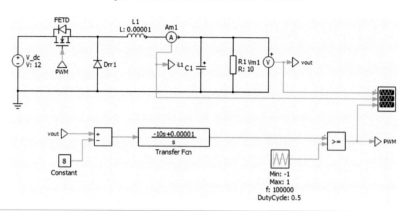

The waveforms of vout, iL1, and PWM are depicted in Figure 6.39.

Figure 6.39: The waveforms of vout, iL1, and PWM.

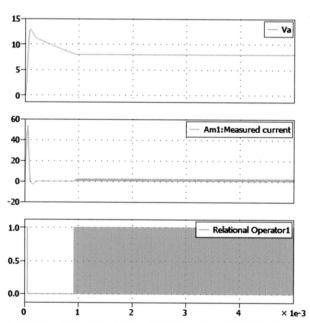

The waveforms of vout, iL1, and PWM. We add ADC A [0 1] and Digital Out, GPIO0, to the simulated circuit in PLECS, as illustrated in Figure 6.40.

Figure 6.40: Adding TI C2000 blocks to the simulated circuit in PLECS.

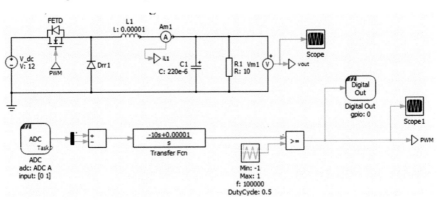

In the Menu, we select the Coder, then in the opened Coder Options window and in the General tab we apply the settings as shown in Figure 6.41.

Figure 6.41: Settings of General tab.

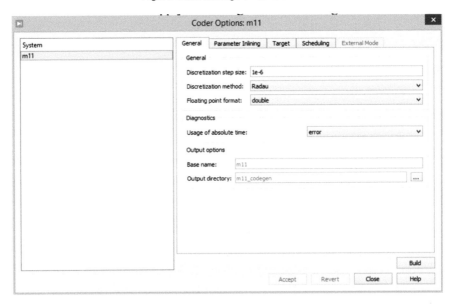

Then, we run the Code Composer Studio (CCS) and, from the Project menu, we select the Import CCS Projects... as depicted in Figure 6.42.

Figure 6.42: Selecting the Import CCS Projects....

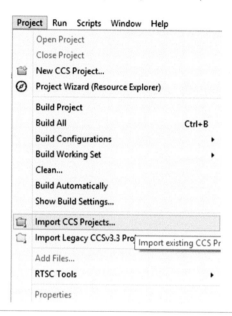

In the installed CCS directory, we click on the installed support packages for TI C2000 folder (tsp_ti_c2000) and then from the projects folder we choose the 28335-zip file and click on the Open button, as illustrated in Figure 6.43.

Figure 6.43: Choosing the 28335-zip file.

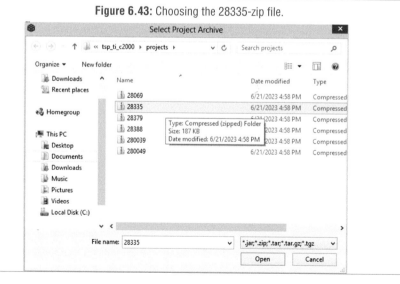

We rename the opened project by right clicking on the opened project and choosing the Rename option, as shown in Figure 6.44.

Figure 6.44: Renaming the opened project.

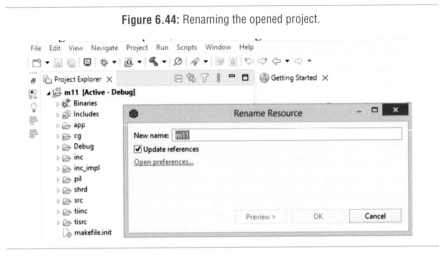

As shown in Figure 6.44, we select the cg folder and right click on it then choose the Properties option and copy the Location section, as you can see in Figure 6.45.

Figure 6.45: Choosing the Properties option and copy the Location section.

In the opened Target tab of Coder Options, we select the TI2833x target and paste the copied "Location" of Figure 6.45 in the CCS project directory and apply the other settings as depicted in Figure 6.46, then click on the Build button.

Figure 6.46: Target settings for the Coder Options.

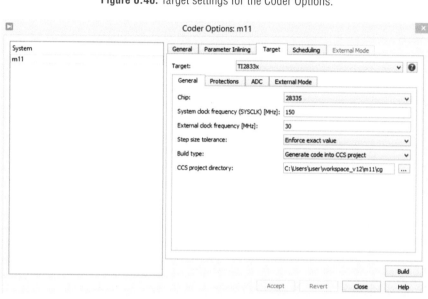

At the end, we select the Project menu and choose the Build All option to compile and build the project as illustrated in Figure 6.47.

Figure 6.47: Building the project.

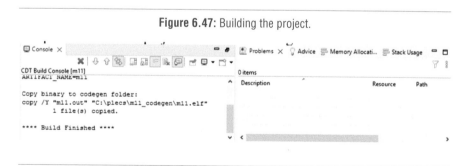

Bibliography

[1]   Texas Instruments, *C2000^{TM} Microcontroller Workshop, Workshop Guide and Lab Manual*, Technical Training Organization, TTO, F28xMcuMdw, Revision 5.0, May 2014.

[2]   M. Patel, The Essential Guide for Developing with C2000^{TM} Real-Time Microcontrollers, 2023.

[3]   Texas Instruments, *C2000 Real-Time Control Peripherals Reference Guide*, 2018.

[4]   Texas Instruments, *Code Composer Studio^{TM} Integrated Development Environment (IDE)*, 2023.

Index

Printed in the United States
by Baker & Taylor Publisher Services